POTENTIAL USE OF
SOLAR ENERGY AND
EMERGING TECHNOLOGIES
IN MICRO IRRIGATION

Innovations and Challenges in Micro Irrigation

VOLUME 4

POTENTIAL USE OF SOLAR ENERGY AND EMERGING TECHNOLOGIES IN MICRO IRRIGATION

Edited by

Megh R. Goyal, PhD, PE
Manoj K. Ghosal, PhD

AAP | APPLE ACADEMIC PRESS

Apple Academic Press Inc. | Apple Academic Press Inc.
3333 Mistwell Crescent | 9 Spinnaker Way
Oakville, ON L6L 0A2 | Waretown, NJ 08758
Canada | USA

© 2017 by Apple Academic Press, Inc.

First issued in paperback 2021

No claim to original U.S. Government works

ISBN-13: 978-1-77463-617-6 (pbk)
ISBN-13: 978-1-77188-364-1 (hbk)

Library and Archives Canada Cataloguing in Publication

Potential use of solar energy and emerging technologies in micro irrigation / edited by Megh R. Goyal, PhD, PE, Manoj K. Ghosal, PhD.

(Innovations and challenges in micro irrigation ; volume 4)
Includes bibliographical references and index.
Issued in print and electronic formats.
ISBN 978-1-77188-364-1 (hardcover).--ISBN 978-1-77188-365-8 (pdf)
1. Microirrigation. 2. Microirrigation--Technological innovations. 3. Photovoltaic power generation. 4. Solar energy. I. Goyal, Megh Raj editor II. Ghosal, M. K., author, editor III. Series: Innovations and challenges in micro irrigation ; v. 4

S619.T74P68 2016   631.5'87   C2016-906162-0   C2016-906163-9

CIP data on file with US Library of Congress

Apple Academic Press also publishes its books in a variety of electronic formats. Some content that appears in print may not be available in electronic format. For information about Apple Academic Press products, visit our website at **www.appleacademicpress.com** and the CRC Press website at **www.crcpress.com**

CONTENTS

LIST OF CONTRIBUTORS

K. Arunadevi, PhD
Assistant Professor, Department of Soil and Water Conservation Engineering, Agricultural Engineering College and Research Institute, Kumulur 621712, Tamil Nadu, India. Email: aruna_swce@yahoo.com

K. Madhava Chandran, PhD
Scientist, Soil Science, Water Management (Agriculture Division), Center for Water Resources Development and Management, Kozhikode 673571, Kerala, India

Vishal K. Chavan, PhD
Assistant Professor and Senior Research Fellow in SWE, AICRP for Dryland Agriculture, Dr. PDKV, Akola, Maharashtra, India. Mobile: +91-7875689897, Email: vchavan2@gmail.com.

Ranjan Kumar Das, PhD
Professor, Department of Farm Machinery and Power, College of Agricultural Engineering and Technology, Orissa University of Agriculture and Technology, Bhubaneswar 751003, Odisha, India

Mohamed El-Sayed El-Hagarey, PhD
Researcher, Irrigation and Drainage Unit, Division of Water Resources and Desert Land, Desert Research Center (DRC), Ministry of Agriculture and Land Reclamation, 1 Mathaf El-Mataria Street, El-Mataria, B.O.P 11753, Cairo, Egypt. Mobile: +201 063031920. Email: elhagarey@gmail.com

M. S. Gaballah, PhD
Researcher, Water Relations Field Irrigation Department, Agricultural and Biological Division, National Research Center, Giza, Cairo, Egypt

M. K. Ghosal, PhD
Professor, Department of Farm Machinery and Power, College of Agricultural Engineering and Technology, Orissa University of Agriculture and Technology, Bhubaneswar 751003, Odisha, India. Email: mkghosal1@rdiffmail.com

H. M. Gitlin, PhD, PE
Retired Extension Specialist in Agricultural Engineering, Cooperative Extension Service, Department of Molecular Biosciences and Bioengineering, University of Hawaii at Mañoa, Honolulu, Hawaii

Megh R. Goyal, PhD, PE
Retired Faculty in Agricultural and Biomedical Engineering, University of Puerto Rico – Mayaguez Campus; and Senior Technical Editor-in-Chief in Agriculture Sciences and Biomedical Engineering, Apple Academic Press Inc., PO Box 86, Rincon PR 00677, USA. Email: goyalmegh@gmail.com

M. Hassib, PhD
Water Relation and Field Irrigation Department, National Research Center, Dokki, Giza, Cairo, Egypt

E. J. Joseph, PhD
Research Scientist, Water Management (Agriculture Division), Center for Water Resources Development and Management, Kozhikode 673571, Kerala, India

Richard Koech, PhD
Professor, School of Environmental and Rural Science, University of New England, Armidale NSW 2351, Australia. Tel.: +61-267735221; Email: rkoech@une.edu.au or richardkoech@hotmail.com.

V. Kumar, PhD

Professor and Head, Department of Agricultural Engineering, Agricultural College and Research Institute, Tamil Nadu Agricultural University, Madurai 625104, Tamil Nadu, India. Email: kumarkncsaga@gmail.com

N. M. Mahrous, PhD

Researcher, Department of Agronomy, Faculty of Agriculture, Cairo University, Cairo, Egypt

Hani A. A. Mansour, PhD

Professor in Soil & Water Eng., Water Relations Field Irrigation Department, Agricultural and Biological Division, National Research Center, Giza, Cairo, Egypt. Email: mansourhani2011@gmail.com

A. M. Michael, PhD

Former Professor and Project Director, Water Technology Center, IARI, New Delhi, Director, IARI, New Delhi; and Ex-Vice Chancellor, Kerala Agricultural University, Trichur, Kerala; Present mailing address: Dr. A. M. Michael, 34/81, V.P. Marikar Road, Edappally North PO, Kochi 682024, Kerala, India

Miguel A. Muñoz-Muñoz, PhD

Ex-President of University of Puerto Rico, University of Puerto Rico, Mayaguez Campus, College of Agriculture Sciences, Call Box 9000, Mayagüez, PR 006819000, USA. Tel.: +1 7872653871, Email: miguel.munoz3@upr.edu

Bernard Omodei, PhD

President, Measured Irrigation, Dot2dot Post Pty Ltd, 5/50 Harvey Street East, Woodville Park, SA 5011, Australia. Email: bomodei@measuredirrigation.com.au

Sabreen Kh. Pibars, PhD

Professor, Water Relations and Field Irrigation Department, Agricultural and Biological Division, National Research Center, Giza, Cairo, Egypt. Email: sabreennrc@yahoo.com

P. K. Selvaraj, PhD

Professor, Agricultural Research Station, TNAU, Bhavanisagar, Tamil Nadu, India

Gajendra Singh, PhD

Ex-President (2010-2012) of ISAE, Former Deputy Director General (Engineering) of ICAR, and Former Vice-President/Dean/Professor and Chairman, Asian Institute of Technology, Thailand; Mailing address: C86 Millennium Apartments, Plot E-10A, Sector 61, Noida 201301, Uttar Pradesh, India. Email: prof.gsingh@gmail.com

Mona A. M. Soliman, PhD

Researcher, Water Relation and Field Irrigation Department, National Research Center, Dokki, Giza, Cairo, Egypt

U. Surendran, PhD

Professor (Soil Science), Water Management (Agriculture Division), Center for Water Resources Development and Management, Kozhikode 673571, Kerala, India. Email: u.surendran@gmail.com

I. Pai Wu, PhD, PE

Retired Professor, Department of Molecular Biosciences and Bioengineering, University of Hawaii at Mañoa, Honolulu, Hawaii; Mailing address: 3050 Maileway, Honolulu, Hawaii 96822, USA. Tel. +1 8089568809. Email: ipaiwu@hawaii.edu

LIST OF ABBREVIATIONS

AD	anaerobic digestion
AFE	annual fixed energy
AIEI	annual total irrigation energy inputs for applying water
ATEI	annual total energy inputs
ATEO	annual total irrigation energy outputs
AWC	available water content
BC	benefit-cost
BD	buried drip irrigation system
CBR	cost benefit ratio
CCS	commercial cane sugar
CDI	conventional drip irrigation
CEC	cation exchange capacity
CLAC	Central Laboratory for Agricultural Climate
CPCRI	Central Plantation Crops Research Institute
CRD	complete randomized design
CSRTI	Central Sericultural Research and Training Institute
CU	coefficient of uniformity
CWRDM	Center for Water Resources Development and Management
DAP	days after planting
EAE	energy-applied efficiency
EECI	efficiency of crop irrigation
EP	energy productivity
ET	evapotranspiration of crop
ETP	potential evapotranspiration
FS	fixed sprinkler irrigation system
FYM	farmyard manure
GI	galvanized iron
GOI	Government of India
gpm	gallons per minute
HDPE	high-density polyethylene
ICAR	Indian Council of Agricultural Research
ICRISAT	International Crops Research Institute for the Semi-Arid Tropics
IDE	International Development Enterprises
IE	energy inputs

IE	installation energy
INCID	Indian National Committee on Irrigation and Drainage
IR	irrigation requirements
IS	irrigation systems
K	potassium
LAI	leaf area index
LCC	life cycle cost
LCDI	low-cost drip irrigation
LDPE	low-density polyethylene
lph	liters per hour
lps	liters per second
MI	measured irrigation
MPPT	maximum power point tracking
N	nitrogen
NCPA	National Committee on the use of Plastics in Agriculture
NEG	net energy gain
NUE	nutrient use efficiency
OE	operating energy
OUAT	Orissa University of Agriculture and Technology
P	phosphorus
PLW	physiological loss in weight
PS	portable sprinkler irrigation system
PV	photovoltaic
PVWPS	photovoltaic water pumping system
RCBD	randomized complete block design
RCE	relative consumed energy
RDF	recommended dose of fertilizer
RDIF	recommended dose of inorganic fertilizers
SD	surface drip irrigation system
SPV	solar photovoltaic
SPV	special purpose vehicle
SPVPS	solar photovoltaic pumping system
SPVWPS	solar photovoltaic water pumping system
TDH	total dynamic head
TNPFP	Tamil Nadu Precision Farming Project
UV	ultra violet
VCE	vermicompost extract
WA	water amounts
WUE	water use efficiency
ZECC	zero energy cool chamber

PREFACE

Crop production suffers mainly from the availability of water, and water is the most limiting factor in the world agricultural scenario. Furthermore, unscientific use of the available irrigation water compounds the problems in crop production. In the present era of acute water shortage caused by over-utilization and depletion of both surface and subterranean water resources, employment of suitable water management practices for effective utilization of available resources in an economic way is of prime importance. It is necessary to develop a proper irrigation scheduling for agricultural crops and to optimize the water and fertilizer requirement under drip fertigation with different crop geometries. Water, N, and K fertigation requirements should be optimized.

Efficient utilization of available water resources is crucial in India, which shares 17% of the global population with only 2.4% of land and 4% of water resources in world. The per capita availability, in terms of average utilizable water resources, which was 6008 m^3 per year in 1947 (presently 1250 m^3) is expected to dwindle to 760 m^3 per year by the year 2050. The food production in India will have to be doubled to 400 million tons, to meet the food security, income and nutritional needs of the projected population in 2020. Agriculture is by far the largest (81 %) water consumer in India. Therefore, more efficient use of water for crop production must be top most priority. The overuse of water in India reflects low irrigation efficiency of about 25–35% in most irrigation systems, with efficiency of 80–95% in micro irrigation systems.

Besides the land, water is also an important factor in the progress of agriculture. In a vast country like India with a geographical area of 328 million hectares, less than 45% area is cultivated. Of this cultivated area, only 35% (65 million ha) is under irrigation. Since water is the limiting factor today, we must utilize it properly to achieve maximum benefits of irrigation. The expansion of area under irrigation is essential for obtaining increased agriculture production required to feed India's growing population of 1202 million. This expansion is possible only through conservation measures, efficient management of the available water resources, and use of efficient irrigation methods. This can be achieved by introducing advanced and sophisticated methods of irrigation, viz. drip irrigation, sprinkler, etc.

Micro/drip/ or trickle irrigation is an application of water at low volume and frequent intervals under low pressure near the root zone of a plant. The drip irrigation is an advanced method of irrigation suitable for water scarcity areas. It uses the water very efficiently with an application efficiency of up to 95%. It applies the correct quantity of water to the plant at the root zone through a network of tubings. The design of a drip irrigation system must be done by an engineer. The system includes design of pipe network, accessories and other equipments, emitters. This method can be very well adopted for wide spaced and closely spaced, ornamental, field crops, vegetable, and fruit crops.

Micro irrigation is also commonly used in crops under greenhouse technology. Thermal modeling of a controlled environment in a greenhouse is required to optimize the various parameters involved in either heating or cooling of greenhouse. The modeling can also be used to optimize greenhouse air temperature (one of the important constituents of the environment inside greenhouse) for enhancing production of a crop from greenhouse for a given thermal capacity. Thermal modeling requires basic energy balance equation for different components of greenhouse system for a given climatic (solar radiation, ambient air temperature, relative humidity, wind velocity, etc.) and design (volume, shape, height, orientation, latitude, etc.) parameters. To facilitate the modeling procedure, a greenhouse can be considered to be composed of a number of separate but interactive components: the greenhouse cover, the floor, the growing medium, enclosed air, and the plant. The crop productivity depends on the proper environment and more specifically on the thermal performance of the system. In this book volume, a mathematical model with suitable assumptions is presented to evaluate thermal performance of a solar greenhouse for water saving and sustainable farming.

In addition, energy demand is growing exponentially in each segment due to the continuous growth and expansion in different sectors like industry, agriculture, irrigation, transportation, communication, housing, health, education, city modernization, entertainment, etc. To meet the increasing demands of energy, the share of coal-based power plants for power generation in India is also rising day by day, causing severe environmental hazards and thus global warming by releasing a considerable amount of greenhouse gases to the atmosphere. The only alternative in this context is to supplement the existing power sector with non-conventional energy sources. Among the non-conventional energy sources, the solar energy appears to be an attractive and viable proposition because of the abundant and free availability of sunshine in the tropical areas.

Moreover, electricity from solar photovoltaic (PV) system is now gaining more importance because of the rapid decline in the cost of solar PV modules through advances in research and development in this area. The attention of planners, policy makers, and researchers is also now diverted to the applications of solar PV system for pumping of water in irrigation sector due to recent increased water demands in agricultural sector and availability of water has become more crucial than ever before.

In developing countries, electrical and diesel-powered water pumping systems are most widely used for irrigation systems. A source of energy to pump water is also a big problem in these countries. Developing a grid system is often too expensive because rural villages are frequently located too far away from existing grid lines. Even if fuel is available, transporting that fuel to remote and rural villages can be difficult. There are not adequate link- roads or supporting infrastructure in many remote villages. The use of renewable energy is therefore of utmost importance for water pumping applications in remote areas of many developing countries. Transportation of renewable energy systems, such as PV pumps, is much easier than that of the other types because they can be transported in pieces and reassembled on site.

PV energy production is recognized as an important part of the future energy generation, because it is non-polluting, free in its availability, and is of high reliability. These facts make the PV energy resource more attractive for many applications, especially in rural and remote areas of the developing countries. Solar PV water pumping has been recognized as suitable for grid-isolated rural locations in places where there are high levels of solar radiation. Solar PV water pumping systems can provide water for irrigation without the need for any kind of fuel or the extensive maintenance as required by diesel and electric pump sets. They are easy to install and operate, highly reliable, durable, and modular, which enable future expansion. They can be installed at the site of use, avoiding the spread of long pipelines and infrastructures.

In this book volume, Bernard Omodei and Richard Koech have discussed a new concept of Measured Irrigation (MI), which is a low pressure micro irrigation system that controls the application rate to each plant. The application rate to each plant is directly proportional to the net evaporation (= evaporation minus rainfall). With measured irrigation, the plants to be irrigated are often grouped into sectors (zones) whereby the irrigation of each sector is independent of all the other sectors. For each sector, the emitters should satisfy the measured irrigation principle which is defined as follows: "For any two emitters in a sector and at the same pressure, the ratio of the flow rates is independent of the pressure within the operational pressure range for the sector". For measured irrigation, an emitter may be a dripper, a length of micro tube, or a nozzle. The term nozzle refers to a short cylindrical tube or hole for restricting the flow. Pressure compensating drippers should not be used for measured irrigation. MI does not require access to an urban water supply or to electricity grid power, and so there are no ongoing costs for reticulated water or electricity. This makes the system particularly suitable to poorer countries, where access to these facilities is either unreliable or too expensive. For conventional pressurized irrigation systems, the volume of water delivered to a plant during the irrigation event depends upon the flow rate. But for MI, the volume of water delivered to a plant during the irrigation event is independent of the flow rate. MI had been successfully implemented in a number of community gardens in Australia.

Since 1971, I have observed micro irrigation systems in almost all vegetable and fruit crops; in urban and home landscaping; in fields for research and farming; in non-automated and fully automated farms; in front of local shops and shopping malls; and for a single plant to 46 million ornamental plants (Dubai Miracle Garden that I personally visited in April 2015). **Dubai Miracle Garden** is located in the North West Quadrant of Arabian Ranches interchange along Shiekh Mohammad Bin Zayed Road within Dubai Land Development Project in Dubai. Miracle Garden has the record in the *Guinness Book of Records* for having the longest flower wall, which gives a landmark for Miracle Garden and City of Dubai. **Dubai Miracle Garden** under micro irrigation is a fully automated system (see image below). Drip irrigation technology has changed drastically since 1970. Now this technology is even mobile-friendly. WOW... I am surprised!

Photo courtesy: <http://www.dubaimiraclegarden.com/panorama/>

The mission of this book volume is to serve as a reference manual for graduate and undergraduate students of agricultural, biological, and civil engineering and horticulture, soil science, crop science, and agronomy. I hope that it will be a valuable reference for professionals who work with micro irrigation and water management; for professional training institutes, technical agricultural centers, irrigation centers, Agricultural Extension Services; and other agencies that work with micro irrigation programs. I cannot guarantee the information in this book series will be enough for all situations.

After my first textbook, *Drip/Trickle or Micro Irrigation Management* by Apple Academic Press Inc., and response from international readers, I was motivated to bring out for the world community this ten-volume series on **Research Advances in Sustainable Micro Irrigation.** During 2014–2015, Apple Academic Press Inc. has published for the world community the ten-volume series on **Research Advances in Sustainable Micro Irrigation,** edited by M. R. Goyal. The website <appleacademicpress.com> gives details on these ten book volumes.

This book is volume four in the book series *Innovations and Challenges in Micro Irrigation*. These book series are a must for those interested in irrigation planning and management, namely, researchers, scientists, educators and students.

The contributions by the cooperating authors to this book series have been most valuable in the compilation of this volume. Their names are mentioned in each chapter and in the list of contributors. This book would not have been written without the valuable cooperation of these investigators, many of whom are renowned scientists who have worked in the field of micro irrigation throughout their professional careers. I am glad to introduce Dr. Manoj K. Ghosal, who is a full Professor in the Department of Farm Machinery and Power, College of Agricultural Engineering and Technology, at Orissa University of Agriculture and Technology. Without his support and extraordinary work on solar energy applications in agriculture, readers would not have this quality publication.

I would like to thank editorial staff, Sandy Jones Sickels, Vice President, and Ashish Kumar, Publisher and President at Apple Academic Press, Inc., for making every effort to publish the book when the diminishing water resources are a major issue worldwide. Special thanks are due to the AAP Production Staff for the quality production of this book. We request the reader to offer us your constructive suggestions that may help to improve the next edition.

I express my deep admiration to my family for understanding and collaboration during the preparation of this book, especially my wife, Subhadra

Devi Goyal. As an educator, there is a piece of advice to one and all in the world: "Permit that our almighty God, our Creator and excellent Teacher, irrigate the life with His Grace of rain trickle by trickle, because our life must continue trickling on..."

—Megh R. Goyal, PhD, PE
Senior Editor-in-Chief
August 01, 2016

WARNING/DISCLAIMER
USER MUST READ IT CAREFULLY

The goal of this compendium, ***Potential of Solar Energy and Emerging Technologies in Sustainable Micro Irrigation,*** is to guide the world engineering community on how to efficiently design for economical crop production. The reader must be aware that the dedication, commitment, honesty, and sincerity are most important factors in a dynamic manner for a complete success.

The editors, the contributing authors, the publisher, and the printer have made every effort to make this book as complete and as accurate as possible. However, there still may be grammatical errors or mistakes in the content or typography. Therefore, the contents in this book should be considered as a general guide and not a complete solution to address any specific situation in irrigation. For example, one size of irrigation pump does not fit all sizes of agricultural land and to all crops.

The editors, the contributing authors, the publisher, and the printer shall have neither liability nor responsibility to any person, any organization or entity with respect to any loss or damage caused, or alleged to have caused, directly or indirectly, by information or advice contained in this book. Therefore, the purchaser/reader must assume full responsibility for the use of the book or the information therein.

The mentioning of commercial brands and trade names are only for technical purposes. It does not mean that a particular product is endorsed over to another product or equipment not mentioned. Author, cooperating authors, educational institutions, and the publisher Apple Academic Press Inc. do not have any preference for a particular product.

All weblinks that are mentioned in this book were active on December 31, 2015. The editors, the contributing authors, the publisher, and the printing company shall have neither liability nor responsibility, if any of the weblinks is inactive at the time of reading of this book.

ABOUT SENIOR EDITOR-IN-CHIEF

Megh R. Goyal, PhD, PE, is a Retired Professor in Agricultural and Biomedical Engineering from the General Engineering Department in the College of Engineering at University of Puerto Rico–Mayaguez Campus; and Senior Acquisitions Editor and Senior Technical Editor-in-Chief in Agriculture and Biomedical Engineering for Apple Academic Press Inc.

He has worked as a Soil Conservation Inspector and as a Research Assistant at Haryana Agricultural University and Ohio State University. He was first agricultural engineer to receive the professional license in Agricultural Engineering in 1986 from College of Engineers and Surveyors of Puerto Rico. On September 16, 2005, he was proclaimed as "Father of Irrigation Engineering in Puerto Rico for the twentieth century" by the ASABE, Puerto Rico Section, for his pioneer work on micro irrigation, evapotranspiration, agroclimatology, and soil and water engineering. During his professional career of 45 years, he has received many prestigious awards. A prolific author and editor, he has written more than 200 journal articles and textbooks and has edited over 35 books. He received his BSc degree in engineering from Punjab Agricultural University, Ludhiana, India; his MSc and PhD degrees from Ohio State University, Columbus; and his Master of Divinity degree from Puerto Rico Evangelical Seminary, Hato Rey, Puerto Rico, USA.

ABOUT COEDITOR

Manoj Kumar Ghosal, PhD, is a full Professor in the Department of Farm Machinery and Power at the College of Agricultural Engineering and Technology at Orissa University of Agriculture and Technology, Bhubaneswar, Odisha, India. Dr. Ghosal has authored books, manuals, and book chapters and has published over 95 research papers in peer-reviewed journals as well as more than 40 popular technical articles. His books include *Renewable Energy Resources: Basic Principles and Applications, Fundamentals of Renewable Energy Sources, Farm Power*, and *Renewable Energy Technologies.*

In addition, he has published on earth air heat exchangers, thermal modeling of greenhouses, solar photovoltaic cool chambers, among other topics. He is at present associate editor of the *International Journal of Agricultural Engineering* and the *International Journal of Tropical Agriculture*, and he is currently serving on the editorial boards of a number of national/international scientific/technical journals. He acts as a referee of many national/international journals and as a consultant and resource person to different governmental and nongovernmental organizations. Dr. Ghosal is a member of several state and national level committees working on renewable energy applications and environmental impacts.

He received his BTech degree in agricultural engineering in 1988 from Orissa University of Agriculture and Technology, his MTech degree in 1990 from the Indian Institute of Technology (IIT), Kharagpur, West Bengal, India, and his PhD degree in 2004 from Indian Institute of Technology (IIT), Delhi, New Delhi, India.

ENDORSEMENTS FOR THIS BOOK VOLUME

This textbook explores potential use of solar energy in micro irrigation that can minimize problems of water scarcity worldwide. *The Father of Irrigation Engineering in Puerto Rico of 21st Century and World pioneer on micro irrigation*, Dr. Goyal [my longtime colleague] and his colleagues have done an extraordinary job in the presentation of the subject matter.

—Miguel A Muñoz, PhD

I believe that the chapters on emerging technologies in micro irrigation will aid in the food security throughout the world.

—A. M. Michael, PhD

In providing these resources on solar energy and technological advances in micro irrigation, the editors as well as the Apple Academic Press are rendering an important service to alleviate energy and water crises.

—Gajendra Singh, PhD

OTHER BOOKS ON MICRO IRRIGATION TECHNOLOGY BY APPLE ACADEMIC PRESS, INC.

Management of Drip/Trickle or Micro Irrigation
Megh R. Goyal, PhD, P.E., Senior Editor-in-Chief

Evapotranspiration: Principles and Applications for Water Management
Megh R. Goyal, PhD, P.E., and Eric W. Harmsen, Editors

Book Series: Research Advances in Sustainable Micro Irrigation
Senior Editor-in-Chief: Megh R. Goyal, PhD, PE

Book Series: Innovations and Challenges in Micro Irrigation
Senior Editor-in-Chief: Megh R. Goyal, PhD, PE

PART I
Basics of Micro Irrigation

CHAPTER 1

WATER AND NUTRIENT MANAGEMENT IN DRIP IRRIGATION IN INDIA: REVIEW

U. SURENDRAN[1], V. KUMAR[2], K. MADHAVA CHANDRAN[1], and E. J. JOSEPH[1]

[1]Water Management (Agriculture Division), Center for Water Resources Development and Management, Kozhikode 673571, Kerala, India

[2]Department of Agricultural Engineering, Agricultural College and Research Institute, Tamil Nadu Agricultural University, Madurai 625104, Tamil Nadu, India

CONTENTS

ABSTRACT

Significant water shortage is being experienced in many countries, particularly in India. Water for agriculture is becoming increasingly scarce in the light of growing water demands from different sectors. Water use per unit irrigated area will have to be reduced in response to limitations in water availability and other associated environmental and societal problems. One of the scientifically proven ways to reduce the total water required for irrigation is to adopt drip irrigation, which can improve crop yield per unit volume of water used (water productivity). Drip irrigation is an efficient method of providing water directly to the root zone of plants, minimizing conventional losses such as deep percolation, runoff, and soil erosion. Drip irrigation was introduced in India for commercial adoption in early seventies and its growth has gained momentum in the last few years only, primarily due to the subsidy extended by Central and State Governments.

Even though drip irrigation is scientifically proven technology for improvement in water and crop productivity, the adoption level is not at the expected/desired level. Adoption rate of drip irrigation is lower than what is predicted due to the difficulties associated with the ecological and socio-economic constraints that exist in India. The constraints which determine the drip adoption and also the major factors influencing farmers' adoption decisions on drip irrigation are reviewed. Based on reviewing the literature, we could identify that the focus on promotion of drip irrigation needs to be shifted from a water-saving technology to improved productivity with less water and nutrients. It has been felt that so far, the benefits, as communicated by the extension officials are mainly in terms of water saving only and not on productivity basis. Since farmers have been getting water at low cost/no cost from the public irrigation system and well irrigation, there is less incentive for them to adopt this capital intensive technology, unless it becomes absolutely necessary. Hence, thrust should be provided on promoting the drip fertigation as a technology to farmers for achieving the higher productivity and profitability on a sustained basis. In this chapter, crop water requirement of major crops was computed with FAO CROPWAT for humid and semi-arid regions India. This has given an insight in to the impact of climatic variability on water requirement of crops. Besides, a detailed description about drip fertigation viz., fertigation units, selection of suitable fertilizers by considering the compatibility, nutrient requirement of crops, and its impact on several crops are discussed in this chapter. A wide spectrum of climate change scenarios has also been discussed in the chapter along with the guidelines for future management of water and nutrients. Recommendations are also provided that will help in developing policy and institutional interventions to encourage adoption of drip fertigation technologies in these regions for improved productivity.

1.1 INTRODUCTION

Efficient utilization of available water resources is crucial in India, which shares 17% of the global population with only 2.4% of land and 4% of water resources in world. The per capita availability, in terms of average utilizable water resources, which was 6008 m^3 per year in 1947 (presently 1250 m^3) is expected to dwindle to 760 m^3 per year by the year 2050. The food production in India will have to be doubled to 400 million tons, to meet the food security, income, and nutritional needs of the projected population in 2020 [38]. Agriculture is by far the largest (81 %) water consumer in India. Therefore, more efficient use of water for crop production must be the topmost priority [51, 54].

The overuse of water in India reflects low irrigation efficiency of about 25–35% in most irrigation systems, with efficiency of 40–45% in a few exceptional cases. Study reveals that Krishna, Godavari, Cauvery, and Mahanadi rivers or irrigation systems have a very low efficiency of around 27%, while Indus and Ganga are doing better with efficiencies ranging from 43 to 47% [7, 8, 26]. This is understandable, since the peninsular rivers have large areas under irrigation in delta areas, where the water management practices are poor, while rotational water supply is practiced in Indus and Ganga systems. According to a study sponsored by Food and Agriculture Organization (FAO), it is estimated that on an average, overall water use efficiency (WUE) of irrigation in developing countries is only about 38%. In the irrigation sector, this would mean more productive and efficient use of the water, that is, "more crop per drop." A report of the task force on irrigation by planning commission, Govt. of India [8, 9] recommends the following water policies:

- Irrigation efficiency needs to be brought at par with international standards: 60% in surface water.
- Irrigation efficiency with ground water should be targeted to 80–85%.
- Irrigation water management should be the utmost priority. Water users associations should be formed and promulgated.

Significant water shortage is being experienced in many countries, particularly in India. Water for agriculture is becoming increasingly scarce in the light of growing water demands from different sectors. By considering all these factors, water use per unit irrigated area will have to be reduced in response to limitations in water availability and other associated environmental and societal problems. One of the scientifically proven ways to reduce the total water required for irrigation is to adopt water-saving method of drip irrigation, which can improve crop yield per unit volume of water used (water productivity). Drip irrigation is an efficient method of providing water directly to the root zone of plants, minimizing conventional losses such as deep percolation, runoff, and soil erosion [10, 11]. Drip irrigation was introduced in India for commercial adoption in early seventies and its growth has gained momentum in the last few years only, primarily due to the subsidy extended by Central and State Governments. From 1500 ha in 1985, the area under drip irrigation in India has increased to 1.9 million ha in 2010 [13]. Presently, India stands first in the world in area under drip irrigation.

Large chunk of money has been provided in the form of subsidy to farmers for installing micro irrigation methods including drip irrigation by Government agencies in India. It has been proved over time and space that drip irrigation can result in more than 50% saving in water with high levels of water use efficiencies for a wide range of crops [10]. It has been scientifically recognized that adoption of drip fertigation method is an option for efficient use of water and nutrients through improvement in crop yield per unit volume of water and nutrients used [2, 10, 38]. Drip irrigation method is well suited for widely spaced high value row crops. The required quantity of water is provided to each plant daily at the root zone through a network of pipe systems. Hence, there is no loss of water either in conveyance or in distribution. Evaporation losses from the soil surface are also very little, since water is applied only to the root zone, and crop canopy provides shade to prevent evaporation. Research studies have indicated that the water saving is about 12–81 % and the yield is increased by 2–179% for various crops, when drip irrigation is used (Table 1.1). Details of increase in yield and water saving under drip irrigation with various crops have been compiled from various sources (Table 1.1).

In this chapter:

- Primary objective was to analyze the challenges in adoption of micro irrigation by farmers and suggesting the ways and means for improving the adoption of drip irrigation [28, 33].
- Secondary objective was to give recommendations on water and nutrient requirement of crops under drip irrigation and fertigation.

TABLE 1.1 Water Saving and Productivity Gain through Drip Irrigation.

Crop	Water consumption (mm/ha)		Yield (Tons/ha)		Water saving over FI (%)	Yield increase over FI (%)	Water use efficiency2	
	FI	DI	FI	DI	–	–	FI	DI
Vegetables								
Ash gourd	840	740	10.84	12.03	12	12	77.49	61.51
Beet root	857	177	4.57	4.89	79	7	187.53	36.20
Bottle gourd	840	740	38.01	55.79	12	47	22.09	13.26
Brinjal	900	420	28.00	32.00	53	14	32.14	13.13
Cabbage	660	267	19.58	20.00	60	2	33.71	13.35
Cauliflower	389	255	8.33	11.59	34	39	46.67	22.00
Chili	1097	417	4.23	6.09	62	44	259.34	68.47
Oka	535	86	10.00	11.31	84	13	53.50	7.60
Onion	602	451	9.30	12.20	25	31	64.73	36.97
Potato	200	200	23.57	34.42	–	46	8.49	5.81
Radish	464	108	1.05	1.19	77	13	441.90	90.76
Ridge gourd	420	172	17.13	20	59	17	24.52	8.60
Sweet potato	631	252	4.24	5.89	61	40	148.82	42.78
Tomato	498	107	6.18	8.87	79	43	80.58	12.06
Fruit crops								
Banana	1760	970	57.50	87.50	45	52	30.61	11.09
Grapes	532	278	26.40	32.50	48	23	20.15	8.55
Lemon	42	8	1.88	2.52	81	35	22.34	3.17
Papaya	2285	734	13.00	23	68	77	175.77	31.91
Pomegranate[1]	1440	785	55.00	109	45	98	26.18	7.20
Sweet lime[1]	1660	640	100.0	150	61	50	16.60	4.27
Watermelon	800	800	29.47	88.23	Nil	179	27.15	9.07
Other crops								
Coconut	–	–	–	–	60	12	–	–
Cotton	856	302	2.60	3.26	60	25	329.23	92.64
Groundnut	500	300	1.71	2.84	40	66	292.40	105.63
Sugarcane	2150	940	128	170	65	33	16.79	5.53

Note: [*]FI = flood irrigation; [**]DI = drip irrigation; [1]Yield in 1000 numbers; [2]Water consumption (mm) per ton of yield.

Source: INCID [14]; Narayanamoorthy [29–33], Kumar [19–21] and Sivanappan [45–47].

1.2   MISSING LINK IN DRIP IRRIGATION

With increasing demand for limited water resources and need to minimize adverse environmental consequences of irrigation, drip irrigation technology will undoubtedly play an important role in Indian agriculture. Even though drip irrigation is scientifically proven technology for improvement in water and crop productivity, the adoption level is not up to the expected level. Adoption rate of drip irrigation is lower than what is predicted due to the difficulties associated with the ecological and socioeconomic constraints that exist in India. The constraints which determine the drip adoption and also the major factors influencing farmers' adoption decisions on drip irrigation have been reviewed. Based on this, we could identify that the focus on promotion of drip irrigation needs to be shifted from a water-saving technology alone to improved productivity with use of less water and nutrients. It has been felt that so far, the benefits, as communicated by the extension officials, are mainly in terms of water saving only and not on productivity basis. Since farmers have been getting water at low cost/no cost from the public irrigation systems and tube-well irrigation, there is less incentive for them to adopt this capital-intensive technology, unless it becomes absolutely necessary.

With respect to crop productivity, it becomes imperative to adopt drip fertigation technology to achieve the desired productivity. Hence, thrust should be provided on promoting drip fertigation among farmers for achieving higher productivity and profitability on a sustained basis.

With this background, this chapter is formulated about drip fertigation, its methodologies, advantages, disadvantages/constraints, results obtained from drip fertigation, and choice of crops for adopting drip fertigation. Besides, recommendations are also provided that will help in developing policy and institutional interventions to encourage adoption of drip fertigation technology in these regions for improving productivity

1.3   FERTIGATION

Fertigation is the process by which, both water and fertilizers are delivered to the crop simultaneously through drip irrigation system [10, 11]. Fertigation ensures that essential nutrients are supplied precisely at the area of most intensive root activity according to the specific needs of crops and soil type, resulting in higher crop yield and quality of the produce. It is a fact that as long as phosphorus (P) is placed adequately, both nitrogen (N) and potassium (K) can be fertigated.

Fertigation was first developed for field and horticultural crops, and later was used on tree plantations. Nowadays, with the advent of hi-tech agriculture in India (especially in Kerala and Tamil Nadu) viz., polyhouses, precision farming, and greenhouse cultivation, the use of fertigation with automatic scheduling of irrigation cycle for agriculture purpose is gaining importance in these parts of India. Today, fertigation is used in any system, small or large scale, all over the world.

The shortage of water worldwide for use in agriculture and increased urbanization have forced agricultural development to new locations, less suitable to old flood or canal irrigation methods. In arid areas, the shortage of potable water and increase of population is driving agricultural growers to use such type of water-saving technologies, whereas such a situation does not exist in semi-arid and humid tropical regions. According to several investigators [10, 11], drip fertigation will continue to expand and slowly replace traditional flood irrigation wherever population demand for fresh water will put pressure on water resources. This holds good for semi-arid and humid tropical regions of India since the pressure on water resources are increasing day by day. Labor costs are also an important factor in the transformation from flood or canal irrigation to permanent fertigation systems. As farmers have already shifted from subsistence to profitable agriculture, the shift to fertigation is inevitable.

Fertigation provides a variety of benefits to the users, such as high crop productivity and quality, resource use efficiency (fertilizer, water, land, and energy), environmental safety, flexibility in field operations, effective weed management, and successful crop cultivation on fields with undulating topography. Fertigation is considered eco-friendly, since it avoids the leaching of nutrients especially $N-NO_3$, K, and other basic cations, thereby reducing nutrient pollution. Fertigation is one of most successful ways of water and nutrient application through drip system. Yield advantages have been reported across a wide range of crops under diverse agro-climatic situations.

Another important problem associated with most of the soils is timely availability of nutrients. The reasons for non-availability may be soil reaction (pH), fixation of nutrients, low cation exchange capacity (CEC), leaching, erosion of nutrients in humid tropics etc. These problems may be rectified by applying fertilizers in synchrony with crop demand at the root zone in smaller quantities [42]. Under drip fertigation, it is possible to regulate the quantity of water and nutrients based on the crop demand. The desired economic benefits of drip irrigation are possible in field crops only when proper drip fertigation strategies for nutrient application (4Rs:

Right source, Right rate, Right time, and Right place) are adopted [41]. But, these are seldom practiced, resulting in reduced profitability for high capital investments. Hence, focus should be given to drip fertigation for improved productivity and profitability.

1.3.1 ADVANTAGES OF FERTIGATION

- **Uniform application of fertilizer:** In fertigation, fertilizer is applied along with irrigation water, that is, through dripper. Normally, uniformity in drip irrigation system is above 95% and accordingly, fertilizer application also achieves higher uniformity.
- **Placement in root zone**: Fertigation provides the opportunity to apply fertilizers/chemicals in the root zone only, since it is possible to have a control through drip irrigation system.
- **Quick and convenient method:** Fertigation provides management of time and quality at control unit of drip irrigation. No damage to the crop by root pruning, breakage of leaves, or bending of leaves, as in the case of conventional fertilizer application methods.
- **Higher nutrient use efficiency**: The nutrients supplied through fertigation increase their availability, limit the wastage through leaching below rooting depth, and consequently, improve fertilizer-use efficiency, with saving in fertilizers. Nutrient use efficiency (NUE) by crops is greater under fertigation, compared to that under conventional application of fertilizers to soil.
- **Frequent application is possible**: Fertigation provides an opportunity to apply fertilizer more frequently than conventional methods. However, a mechanical spreader is costly, causes soil compaction, may damage the growing crop, and is always not accurate.
- **Possibility of application in different grades** to suit the stage of crop: The soil and plant system, which requires different types of fertilizer materials during the crop cycle, can be supplied through fertigation more effectively, compared to conventional methods. It is possible to supply nutrients according to the crop developmental phases throughout the season in order to meet the actual nutritional requirements of the crop.
- **Efficient delivery of micronutrients** along with NPK: Fertigation provides an opportunity to mix the required micronutrients along with conventional NPK and it can be applied to soil/plant systems.

Otherwise, soil application of micronutrients is very difficult and usually, foliar spray is being practiced.

- **Less water pollution**: The excessive use of fertilizer through conventional methods leads to leaching of fertilizer material beyond the root zone depth. At a number of locations, it has been observed that it pollutes the surface and groundwater of the area. Fertigation provides an opportunity to prevent these environmental hazards.
- **Higher resource conservation:** Fertigation helps in saving of water, nutrients, energy, labor, and time.
- **Healthy crop growth**: When fertigation is applied through the drip irrigation system, crop foliage can be kept dry, thus avoiding leaf burn and delaying the development of plant pathogens.
- **Cultivation on Marginal lands**: Drip fertigation allows cultivation of crops on marginal lands, where accurate control of water and nutrients in the plant root environment is critical.
- **Good quality**: Since all the required nutrients are supplied through drip fertigation, the crop produce obtained are of good and uniform quality.
- **Extended period of harvest:** In the case of vegetable crops, because of drip fertigation, the period of harvest is extended (under drip fertigation, initial harvest starts 15 days earlier than the normal plants and the final harvest also extends by another 15 days).
- **Use of plastic mulch:** However, when fertigation is combined with the use of plastic cover over crop rows, it can bring extra benefits like:

  a. Reduction in the evaporation losses of water from the soil surface.
  b. Development of salinity on soil surface is delayed.
  c. Prevents weed preponderance and consequent reduction in herbicide use.
  d. Soil temperature is also regulated when clear or reflecting type of plastic sheets is used.

1.3.2 FERTIGATION METHODS

Fertigation is a method of fertilizer application, wherein, fertilizer is combined with irrigation water in the drip irrigation system. Recommended water-soluble fertilizers are mixed with water and passed into the system. In this system, the fertilizer solution is distributed uniformly through irrigation water. The availability of nutrients under this is very high, and hence, the efficiency is more. Liquid fertilizers as well as water-soluble fertilizers are used in drip fertigation, and fertilizer use efficiency can be increased up to 80–90%. This unit delivers nutrients to the plants with water and also applies system maintenance chemicals such as acid, chlorine, or other emitter cleaners. There are different types of fertigation units available. They include injection pump, fertilizer tank, and Venturi units. The details about individual units are explained in the following section:

1.3.2.1   VENTURI

This is a very simple and low cost device (Fig. 1.1). A partial vacuum is created in the system, which allows suction of fertilizers into the irrigation system through venturi action. Vacuum is created by diverting a proportion of water flow from the main line and passing it through a constriction (venturi), which increases the velocity of flow, thus creating a drop in pressure. When pressure drops, the fertilizer solution is sucked into the venturi through a suction pipe from the tank, from where it enters into the irrigation stream. Even though simple, venturi may not mix the fertilizers uniformly with water, which may result in uneven fertilizer distribution in the field. The suction rate of venturi is 30–120 lph. It is easy to handle and is affordable even by small farmers. Venturi is more suitable for small landholdings.

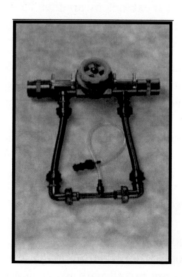

FIGURE 1.1   Venturi.

1.3.2.2   FERTILIZER TANK

A tank containing the fertilizer solution is connected to the irrigation pipe at the supply point (Fig. 1.2). In this system, part of the irrigation water is diverted from the main line to flow through a tank containing the fertilizer in a fluid or granular form. Slight reduction in pressure is created between the off take and return pipes of the tank by means of a pressure-reducing valve, which causes water from the main line to flow through the tank, resulting in dilution and flow of the diluted fertilizer into the irrigation stream. With this system, the concentration of fertilizer entering the irrigation water changes continuously with time, starting from a high to a low concentration. Hence, there will not be uniform fertilizer distribution. Fertilizer tanks are available in 90, 120, and 160 L capacity.

FIGURE 1.2   Fertilizer tank.

1.3.2.3   FERTILIZER PUMP

The fertilizer pump is a standard component of the control head system (Fig. 1.3). The fertilizer solution is held in a non-pressurized tank and it can be injected into the irrigation water at any desired ratio. Therefore, fertilizer availability to each plant is maintained properly. These are piston or diaphragm pumps, which are driven by the water pressure of the irrigation systems such that injection rate is proportional to the flow of water in the system. A high degree of control over the fertilizer injection rate is possible and no serious head losses are incurred. But, the capital cost is high. Another advantage is that if the flow of water stops, fertilizer injection also automatically stops. This is perfect equipment for accurate fertigation. Suction rates of pumps vary from 40 to 160 lph.

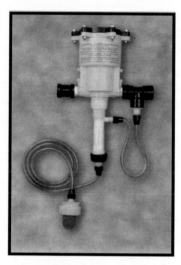

FIGURE 1.3   Fertilizer pump.

A comparative assessment of fertigation equipments has been carried out, based on the feedback got from farmers in the semi-arid region of Tamil Nadu, India, where drip fertigation has been adopted on a considerable scale (Table 1.2).

TABLE 1.2   Comparison of Fertigation Equipment.

Characteristics	Tank	Venturi	Pump
Use of granular/solid fertilizers	Possible	To be dissolved before application	To be dissolved before application
Use of liquid fertilizers	Possible	Possible	Possible
Discharge rate	High	Low	High
Concentration control	None	Medium	Good
Head loss	Low	Very high	Low
Ease of operation	High	Medium	Low
Price	Medium (Rs.4000)	Low (Rs.1500)	High (Rs. 12,000)

Surendran et al. [50].

1.3.3   FERTILIZER APPLICATION IN FERTIGATION

According to Sne [48], to apply the same doses of fertilizers during the specific phenological stage of a plant, two different patterns of application can be made depending on the crop, soil type, and farm management system:

1.3.3.1   QUANTITATIVE DOSING

A measured amount of fertilizer is injected into the irrigation system during each water application. Injection may be initiated and controlled automatically or manually.

1.3.3.2   PROPORTIONAL DOSING

In this process, a constant predetermined ratio between the volume of the irrigation water and the volume of the fertilizer solution is maintained, resulting in a constant nutrient concentration in the irrigation water.

However, the farmers in semi-arid region of Tamil Nadu arc adopting fertigation in the following four ways:

- **Continuous application:** Fertilizer is applied at a constant rate from the beginning to the end of the irrigation cycle. The total amount of fertilizers is injected, regardless of water discharge rate.
- **Three-stage application:** Irrigation starts without fertilizer. Injection begins when the ground is wet. Injection is stopped before the irrigation cycle is completed. Remainder of the irrigation cycle allows the fertilizer to be flushed out of the system for the system cleansing. In simpler terms, suppose the farmer is irrigating for 45 min, then, initial 15 min will be only irrigation to wet the soil, the next 15 min period for fertigation to supply the nutrient, and the final 15 min of irrigation to flush out the nutrients from the drip laterals. This is called middle rule of fertigation: Fertilizer is injected during second 1/3rd of irrigation cycle.
- **Proportional application:** The injection rate is proportional to the water discharge rate, for example, 1 L of fertilizer solution is mixed into 1000 L of irrigation water. This method has the advantage of being extremely simple and allows for increased fertigation during the periods of high water and nutrient demand.
- **Quantitative application:** Nutrient solution is applied in a calculated amount to each irrigation block for example, 20 L to block A and 40 L

to block B. This method is suited in automation and allows the placement of the nutrients by controlling precisely.

1.3.4 SUITABILITY OF FERTILIZERS FOR FERTIGATION

A wide range of fertilizers, both solid and liquid, is suitable for fertigation, depending on the physicochemical properties of the fertilizer solution. For large-scale field operations, solid fertilizer sources are typically less expensive alternative to the commonly used liquid formulations. The solubility of these fertilizers varies greatly. When switching to a solid fertilizer source, problems can be avoided in the nurse tanks by ensuring that ample water is added to the stock solution. Four main factors in selecting fertilizers for fertigation should be considered [18]:

- Plant type and stage of growth.
- Soil conditions.
- Water quality.
- Fertilizer availability and price.

The type of fertilizer for fertigation should be of high quality, high solubility and purity, containing low salt levels and with an acceptable pH, and it must fit in the farm management program. Following are the factors to be considered while selecting fertilizers for fertigation:

- Completely water soluble at field temperature conditions.
- No clogging of filters and emitters.
- Fast dissolution and minimal interaction with irrigation water.
- Minimum content of conditioning agents/filler materials.
- No drastic changes in pH of water.
- Low corrosiveness for control head and system components of drip irrigation.
- Selection of fertilizers should be done based on the water quality results. For example, even good soluble P sources get precipitated if Ca+Mg in water is more than 50 ppm and bicarbonate content in water is more than 150 ppm. Similarly, if water contains iron, it will form an insoluble iron phosphate.
- When applying fertilizers through irrigation water under fertigation, it is essential to be familiar with some important facts regarding solubility of fertilizers.

1.3.5 SOLUBILITY OF FERTILIZER

The solubility of a fertilizer is defined as the maximal amount of the fertilizer that can be completely dissolved in a given amount of distilled water at a given temperature. Upon request, different manufacturers may provide different solubility data of their fertilizers. The average solubility data of various fertilizers are shown in Table 1.3.

TABLE 1.3   Fertilizers Suitable for Drip Fertigation.

Nutrient	Water-soluble fertilizers	Nutrient content (%)	Solubility (g/L)	pH
Nitrogen	Urea	46-0-0*	1100	5.8
	Ammonium nitrate	34-0-0	1920	5.7
	Ammonium sulfate	21-0-0	750	5.5
	Calcium nitrate	16-0-0	1290	5.8
	Magnesium nitrate	11-0-0	720	5.4
	Urea ammonium nitrate	32-0-0		5.7
	Potassium nitrate	13-0-46	316	5.4
	Mono ammonium phosphate	12-61-0	410	4.9
Phosphorus	Mono ammonium phosphate	12-61-0	410	4.9
	Mono potassium phosphate	0-54-32	250	5.5
	Phosphoric acid	0-82-0		2.6
Potassium	Potassium chloride	0-0-60	340	7.0
	Potassium sulfate	0-0-50	110	3.7
	Potassium nitrate	13-0-46	316	5.4
	Potassium thiosulfate	0-0-25	110	3.7
	Mono potassium phosphate	0-52-34	250	5.5
NPK	Polyfeed	19-19-19	750	5
		20-20-20	730	5
Micronutrients	Fe EDTA-Fe	13	120	4.0–5.5
	Fe DTPA-Fe	12	60	7.0–8.0
	Fe EDDHA-Fe	6	110	4.0–5.5
	Zn EDTA-Zn	15	1000	6.0–7.0
	Cu EDTA-Cu	14	1200	6.0–7.0
	Solubor-B	20	210	4.0–5.0
	Combined micro nutrients (B+Cu+Fe+Mn+Mo+Zn+Mg)	**	800	6.5–8.5

*Indicates NPK content.

**7.1% Fe, 3.48% Mn, 1.02% Zn, 0.76% Cu, 0.485% Mo, 0.65 % B.

1.3.6   MIXING FERTILIZERS

When a fertilizer is dissolved in water, it should not exceed its solubility. Otherwise, precipitates may form and clog the irrigation system. Moreover, the nutrients intended to be provided through the solution may not be fully available. When mixing fertilizers that contain a common element (for example potassium nitrate together with potassium sulfate), the solubility of the fertilizers is decreased. In such a case, we cannot consider the fertilizer solubility data shown in Table 1.3 alone. The solubility of the mixture will have to be found out by trial and error.

1.3.7   FERTILIZER COMPATIBILITY

Some fertilizers should not be mixed together in one stock tank because an insoluble salt may form very quickly. An example for such incompatibility is mixing fertilizers that contain calcium with those that contain phosphate or sulfate. In order to avoid unwanted precipitates, "jar test can be performed." In this test, fertilizers are initially mixed in a jar containing the same water used for irrigation.

The fertilizers should be mixed exactly in the same concentration as intended to be used in the stock tanks. If a precipitate forms or if the solution has a "milky" appearance, the test should be repeated with lower concentrations of the fertilizers. Fertilizer compatibility chart is given in Table 1.4. It is desirable to use water-soluble specialty fertilizers for fertigation due to the following features:

- Free from chlorides and sodium.
- No salt build up in the crop root zone.
- Contains 100% plant nutrients.
- Fast acting nitrate nitrogen, soluble phosphorus, and soluble potassium.
- Completely water soluble with any residues.
- Most of the fertilizers are acidic in nature, and hence, no special chemical treatment is required to check emitter plugging.
- Maintains optimum soil pH, contributing to more uptake of nutrients.
- Most of the soluble fertilizers are blended with micronutrients.

TABLE 1.4 Fertilizer Compatibility Chart [52].

Fertilizer	Urea	Ammonium nitrate	Ammonium sulfate	Calcium nitrate	Potassium nitrate	Potassium chloride	Potassium sulfate	Mono ammonium phosphate	Iron, zinc, copper, manganese sulfate	Iron, zinc, copper, manganese chelate	Magnesium sulfate	Phosphoric acid
Urea												
Ammonium nitrate												
Ammonium sulfate				XX	X		X					
Calcium nitrate			XX		X		XX	XX	XX	X	XX	XX
Potassium nitrate			X	X			X					
Potassium chloride							X					
Potassium sulfate			X	XX	X	X			X		X	
Mono ammonium phosphate				XX					XX	X	XX	
Iron, zinc, copper, manganese sulfate				XX			X	XX				
Iron, zinc, copper, manganese chelate				X				X				X
Magnesium sulfate				XX			X	XX				
Phosphoric acid				XX						X		

Cells left blank = fully compatible; X = reduced solubility; XX = incompatible.

1.4   REVIEW OF RESEARCH ADVANCES ON WATER AND NUTRIENT MANAGEMENT IN DRIP IRRIGATION: INDIA

The available literature provides sufficient evidence in favor of increased productivity of several crops due to fertigation [10, 11]. Since only a few studies could be traced on the adoption level of farmers on drip fertigation technology, a review was done on the impact of drip fertigation from research-based trials in semi-arid and humid tropical regions, and is presented in the following section.

Howlett and Velkar [40] evaluated the Tamil Nadu Precision Farming Project (TNPFP) in the Dharmapuri and Krishnagiri districts on about 400 ha of land, in which the core technology is drip fertigation technology. They found that the average yield for tomato, eggplant, and banana were at least 3–12 times higher against national average estimates (17.35, 10.46, and 28.58 tons/ha respectively). This study indicates the potential for increasing yields in "real life situations" with proper adoption of technologies and concludes that this is a successful technology. This project also shows the utility of demonstrations carried out with farmers' involvement, and participation of technical experts in the field of drip fertigation.

Another survey by Monitoring and Evaluation Unit of Government of Tamil Nadu reported that 100% of the respondents were happy with the fertilizer application through drip fertigation. Farmers reported that additional yield, high remuneration, water saving, technical guidance, reduced cost of cultivation, and empowerment of farmers are the major benefits obtained through drip fertigation project [9].

A review of the current literature on the use of fertigation on vegetable production, including the constraints to its adoption by Jat et al. [15] from International Crops Research Institute for the Semi-Arid Tropics (ICRISAT), suggests that to make agriculture sustainable and economically viable, there is need to promote fertigation on a large scale by the concerned stakeholders. The review confirms that yield advantages have been reported across the wide range of crops under diverse agro-climatic situations.

The yield simulations based on the field trials and local conditions also indicate that the potential yield increase from drip fertigation seems promising in many regions, based an average yield response from 10 years under different climatic scenarios [39]. In irrigated horticultural production systems, increased precision in the application of both water and nutrients can potentially be achieved by simultaneous application via fertigation [2, 34]. This has the advantage of synchronizing nutrient supply with plant demand [36], thus enabling reduction in the amount of nutrients applied and reducing environmental impact, besides improving crop productivity [35]. The advantage of fertigation over conventional method of fertilizer application has been emphasized by several workers [24, 25, 44]. Fertigation frequency is one major management variable with drip fertigation. It is often assumed that high frequency fertigation is preferable to less frequent fertigation. Several workers advocated frequent fertigation of crops by low-volume irrigation system.

Under highly weathered P-fixing tropical soils and K-fixing vermiculite soils, drip fertigation allows fertilizers to be applied to smaller soil volume in the active root zone of the crop, which in turn ensures greater nutrient availability in the fertilized zone than would be obtained with a broadcast-incorporated application. The result can be greater nutrient uptake from the applied fertilizer [1, 37]. Muralikrishnasamy et al. [27] reported that drip irrigation at 50% PE + 100% N and K through fertigation recorded highest water use efficiency, water productivity, and water saving in chilies over farmers' practice of surface irrigation (0.9 IW/CPE ratio) + entire NPK as soil application.

In sandy soil with low CEC and K fixation, potassium ions move along with water, and thus, it will be prudent to apply K-fertilizers through drip irrigation in more splits to achieve maximum NUE [12]. Shedeed et al. [43] observed significant increase in growth parameters (plant height, leaf area index (LAI), fruit dry weight, and total dry weight), yield components (number of fruits/plants, mean fruit weight, and fruit yield/plant), and total fruit yield with the application of 100% recommended dose of fertilizer (RDF) through fertigation over furrow and drip irrigation and soil application of fertilizers. Bhakare and Fatkal [3] recorded the benefit cost (B:C) ratio of Rs. 3.30 under 100% RDF applied through water-soluble fertilizers in fertigation, as against Rs. 2.78 in 100% RDF with conventional fertilizer application and surface irrigation. They observed highest FUE when 50% RDF was applied through drip irrigation and found lowest FUE when 100% RDF was applied through conventional fertilizer application method and irrigation water was applied by surface application.

A study from Kerala Agricultural University reported that furrow-irrigated okra showed 54% and 57% lower yield than drip irrigation at 1.0 Epan and fertigated with 120 kg N/ha [4]. In eggplant, higher yields (42.33 t/ha in I crop and 37.90 t/ha in II crop) were recorded in treatment: Drip irrigation at 75% of PE with fertigation of 75% of recommended N and K [22]. A study at Central Plantation Crops Research Institute (CPCRI), Kasargod, Kerala, by Ravi Bhat et al. [42] in arecanut showed that fertigation up to 75% NPK provided a higher NUE than the combination of drip irrigation and soil application of 100% NPK, indicating greater production at lesser application rates. The yield increase with 75% NPK fertigated at 10 days interval was 100% higher than for the control. The 11-year study indicated that adoption of fertigation not only increases productivity, but also ensures higher efficiency of the two most critical inputs, that is, water and nutrients in crop production.

The favorable effect of fertigation has also been reported by CPCRI in Kerala. Fertigation has brought conspicuous results in terms of copra yield of coconut cultivated in coastal sandy soil. Significantly, higher nut and copra yield was observed under drip fertigation at the rate of 100% RDF. This yield was at par with the yield obtained under 50 and 75% of fertilizer application through drip irrigation. The possibility of saving 50% of fertilizer, when applied through drip irrigation, is evident from this study. The soil nutrient status with regard to N, P, and K was also found to be more in drip fertigation soil samples, compared to conventional method of fertilizer application, according to annual report 2010–2011 by CPCRI [<www.cpcri. gov.in>: Indian Council of Agricultural Research-Central Plantation Crops Research Institute (ICAR-CPCRI)].

According to Patel and Rajput [38], the yield of okra under conventional method of fertilization with 100% RDFs and under fertigation with 60% RDFs was not significantly different (23.0 t/ha and 23.1 t/ha in the year 2000 and 23.56 t/ha and 23.35 t/ha in the year 2001). This indicates that a saving of 40% in fertilizer use may be achieved if applied through fertigation without affecting the okra yield. More than 16% increase in yield under fertigation (25.21% in the year 2000 and 16.59% in the year 2001) was observed as compared to broadcasting method of fertilizer application when 100% RDFs was applied. Similar results of increase in productivity of chili crop due to fertigation were reported by Veerana et al. [53].

Darwish et al. [5] reported that fertigation with continuous N feeding through drip system based on actual N demand and available N in the soil resulted in 55% N recovery; for spring potato crop in this treatment, 44.8% N need was met from the soil N and 21.8% from the irrigation water. Higher N input increased not only the N derived from fertilizers, but also the residual soil N.

Mahajan and Singh [23] found that in greenhouse-grown tomato, when the same quantity of water and N was applied through drip irrigation, a significantly higher tomato yield (68.5 t/ha) was obtained as compared to the yield of 58.4 t/ha and 43.1 t/ha in check basin method of irrigation when the crop was sown both inside and outside the greenhouse, respectively.

Results obtained from the experiments conducted at Trichy and Karur districts of Tamil Nadu with precision farming clusters laid out in farmers' fields are discussed here. In these experiments, all agronomic practices were carried out as per farmers' practice. The hypothesis tested was—drip fertigation versus conventional soil application of fertilizers and irrigation. It can be seen from Table 1.5 that drip irrigation gives higher yield components and crop yield (43.8%) in tomato than surface irrigation. The data for other crops also showed that there is a considerable improvement in yield drip fertigation, when compared to soil application of fertilizers with flood irrigation (Table 1.6). The influence of drip fertigation in tomato and chilies is evident from the size, uniformity, and quality of the produce (data not shown).

TABLE 1.5   Yield and Yield Components of Tomato under Drip Fertigation.

Details	Number of fruits per plant	Mean fruit weight (g)	Total fruit yield (t/ha)
Surface irrigation with conventional soil application of fertilizers	12.10	84.5	24.03
Drip fertigation	15.40	116.0	34.56

TABLE 1.6   Yield of Selected Crops under Drip Fertigation.

Crop	Location	Yield (ton/ha)		Yield increase (%)
		Surface irrigation with conventional soil application of fertilizers	Drip fertigation	
Banana	Karur	57.50	87.50	52.2
Chilies	Karur	0.75	1.20	60.0
Eggplant	Trichy	24.00	34.10	42.1
Okra	Karur	7.50	10.24	36.5

Source: Drip irrigation—Report No. 28 from EID Parry [50].

Proper fertigation management requires knowledge of soil fertility status. Hence, soil fertility status of the above trials was monitored (Tables 1.5 and 1.6). The results indicated that soil-available N, P, K, and organic carbon after harvest of the crop were significantly higher in drip fertigation plots than the surface-irrigated plots. Drip fertigation significantly improved the soil organic carbon status in the soil and maintained the level to that of initial status as shown in Figure 1.4. This can be attributed to higher root dry matter production and *in situ* root decay in the rhizosphere area due to drip fertigation. It can also be concluded from this figure that under traditional method of P fertilizer application, P fixation happens in the soil, making it unavailable to plants. This problem is overcome, when the fertilizers are applied frequently through drip irrigation in small doses.

The Figure 1.5 reveals that drip fertigation places nutrients in the active root zone, besides maintaining favorable soil moisture content. This results in much greater availability of N and K in the crop root zone due to greater mobility and uptake of nutrients. Further, drip fertigation has the potential to minimize leaching loss and to improve the available N and K status in the root zone. This is confirmed from the graph shown in Figure 1.6 that under conventional method of fertilizer application along with surface irrigation, higher levels of nitrate nitrogen (NO_3-N) were observed at soil depth of 30 cm, when compared to drip fertigation.

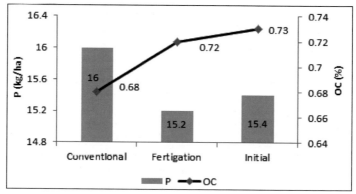

FIGURE 1.4 Soil-available P and organic carbon levels under drip fertigation and conventional fertilizer application under surface irrigation.

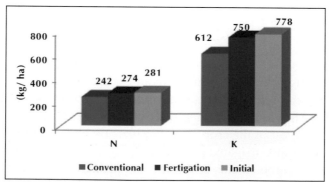

FIGURE 1.5 Soil-available N and K under drip fertigation and conventional fertilizer application under surface irrigation.

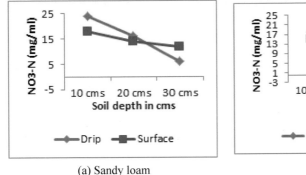
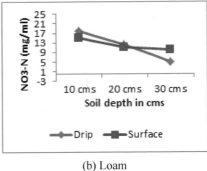

(a) Sandy loam (b) Loam

FIGURE 1.6 Soil leachate nitrate concentration under drip fertigation and conventional fertilizer application under surface irrigation.

A field experiment was conducted to evaluate different irrigation methods (viz., low cost drip irrigation (LCDI), conventional drip irrigation (CDI) with single and paired row, and siphon and flood irrigation) on sugarcane production [6] with and without fertigation. The results showed that cane yield and water productivity was significantly increased in both plant and ratoon crop of sugarcane owing to the methods of irrigation. Among the irrigation methods, LCDI recorded118.6 t/ha of cane yield and it was on par with the single row CDI, which recorded the highest mean yield of 120.4 t/ha and both are found to be significantly superior to the rest of the treatments. The lowest yield was recorded in the treatment of flood irrigation (94.40 t/ha) without drip fertigation (Table 1.7).

Similarly, farmer participatory demonstrations were also conducted in Trichy and Karur districts to demonstrate the impact of drip fertigation to farmers. The yield and quality data showed that there was considerable difference between flood irrigation with soil application of fertilizers and drip-fertigated plots and is statistically significant in all the soil types and locations (Tables 1.8 and 1.9; Fig. 1.7).

In a drip fertigation experiment in banana, subsurface drip fertigation of 100% RDFs (50% P and K as basal, remaining N, P, and K as water-soluble fertilizers) plus liquid bio-fertilizers and subsurface drip fertigation of 100% RDFs (Urea, 13:40:13,KNO3)+ liquid bio-fertilizers were equally effective in increasing growth and physiological parameters of banana. The highest bunch yield, quality parameters, and water use efficiency of banana were recorded in subsurface drip fertigation of 100% RDFs (Urea,

13:40:13,KNO3)+liquid bio fertilizers compared to surface irrigation with soil application of RDFs [10].

TABLE 1.7   Effects of Different Methods of Irrigation on Cane Yield (t/ha).

Treatments	Plant cane yield, t/ha			Ratoon cane yield, t/ha		
	V1-Co 86032	V2-PI 96-0843	Mean	V1-Co 86032	V2-PI 96-0843	Mean
T_1	110.3	108.6	109.4	104.4	103.7	104.1
T_2	124.7	130.3	127.5	119.8	121.0	120.4
T_3	129.5	118.2	123.8	125.3	111.9	118.6
T_4	108.5	116.5	112.5	104.2	108.5	109.3
T_5	93.2	92.5	92.9	95.3	93.5	94.4
Mean	113.2	112.8	113.0	109.8	107.7	109.4
	CD (p = 0.05)	CV%		CD (p = 0.05)		CV%
Irrigation	3.32	7.12		3.16		4.56
Variety	NS			NS		

Source: Drip irrigation—Report No. 28 from EID Parry [50].

TABLE 1.8   Yield Data (t/ha) Obtained from Farmers' Demonstration Plots: *Mean Data of Five Replications.*

Division	Farmer's name	Variety	LCDI	Flood	t test (p = 0.05)
Manapparai	Mr. Chinnakannu	Co 86032	105.3	94.9	3.46
Krishnarayapuram	Mr.Aruvappu	Co 86032	97.0	78.0	3.32
Kulithalai	Mr.Shanmugham	PI 95-0151	118.7	92.4	3.18
Pettavaithalai	Mr.Anbalagan	PI 95-0151	130.0	102.7	3.85
Marudhur	Mr.Theelapan	PI 96-0843	132.5	105.0	4.89
Trichy	Mrs.Sudhadevi	Co 86032	126.4	108.4	3.42

Source: Drip irrigation—Report No. 28 from EID Parry [50].

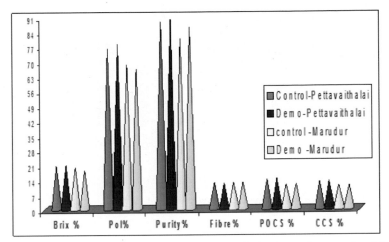

FIGURE 1.7   Sugarcane quality parameters under drip fertigation and conventional fertilizer application.

TABLE 1.9   Quality Parameter, Sugar CCS (%).

Quality parameter	Pettavaithalai		Marudhur	
	Control-flood	Demo-drip	Control-flood	Demo-drip
Commercial cane sugar (CCS, %)	13.07	13.62	11.16	11.58

A greenhouse field experiment was conducted at Center for Water Resources Development and Management (CWRDM), Kozhikode, Kerala, under humid tropical conditions from January 2014 to April 2014 to evaluate the effect of different methods of irrigation and nutrient levels in okra. The study showed that 100% of recommended dose of inorganic fertilizers (RDIF) through surface drip irrigation registered the highest yield followed by 100% of RDIF through subsurface drip irrigation (Fig. 1.8). The yield improvement was 142% higher when compared to the surface flood (channel) irrigation method. In all the drip fertigation treatments, the moisture content was maintained, since the water is being supplied continuously at regular intervals; in flood irrigation, the major factor negatively affecting crop yield, is due to the moisture stress. Increase in nutrient levels improved the okra yield which has been confirmed by the polynomial regression with the R^2 value of 0.98 (Fig. 1.9a and 1.9b).

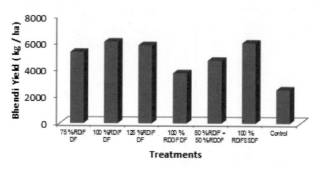

SEd = ±117.8; CD (p ≤ 0.05) = 241.2

FIGURE 1.8   Effect of drip fertigation systems on okra yield.

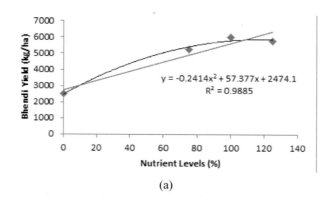

(a)

(b)

FIGURE 1.9   Polynomial regression curve for increase in nutrient levels with okra yield.

In another study done among Palakkad farmers who have adopted fertiga-tion using subsidy of Government, it was observed that invariably, in all the plots, the drip-fertigated area performed better than the control with respect to maintaining soil moisture and soil fertility status (Fig. 1.10 and 1.11). Soil organic carbon and soil-available potassium were high in drip-fertigated plots when compared to surface-irrigated plots. In drip-fertigated plots, the soil moisture content was always higher than available water content (AWC) of 12%, whereas, in control plot, it is lower than AWC (Fig. 1.12). The soil-wetting pattern under drip fertigation plot of cluster bean in Palakkad district shows that soil moisture is maintained up to 45 cm soil depth and also from 45 cm distance from the emitter. This implies that horizontal movement of moisture is taking place and hence, maintaining the adequate moisture level is required for effective plant growth.

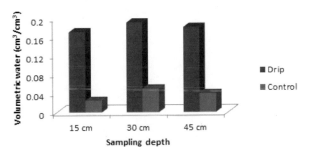

FIGURE 1.10   Soil moisture content of drip fertigation and control plot in Palakkad district.

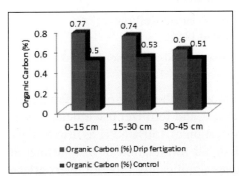
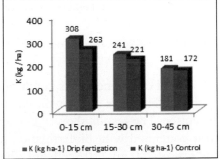

FIGURE 1.11   Impact of drip fertigation on soil organic carbon and potassium in Cluster bean plot: Volumetric water content (cm³/cm³).

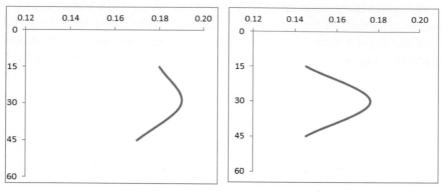

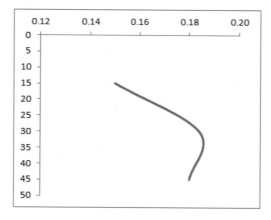

(c) 45 cm from the emitter

All figures: Volumetric water content (cm³/cm³)
Note: AWC: about 12 %

FIGURE 1.12 Soil moisture content of drip fertigation and control plot in Palakkad district.

A field experiment was conducted to assess the effect of drip fertigation with bio-fertilizers on growth, yield, water, and fertilizer use efficiencies of Bt cotton. The results showed that application of 150% NPK as drip fertigation combined with biofertigation of liquid formulation of azophosmet at the rate of 250 mL (10^{12} cells/mL)/ha was significantly superior over surface irrigation and soil application of 100% NPK and registered the highest seed cotton yield of 3395 kg/ha. The soil fertility status was superior under this treatment when compared to surface irrigation and soil application of fertilizers. In this study, the authors recommended that drip fertigation of inorganic fertilizers in combination with biofertigation was the viable agro technique to realize the yield potential of Bt cotton and sustenance of soil fertility [16, 17].

A five-year field trial conducted on a laterite soil to evaluate the effects of organic and inorganic fertigations in arecanut-sole and arecanut–cocoa land use systems showed that fertigation of 75% NPK, vermicompost extract (VCE) 20% N, and VCE (10 and 20% N) 25% NPK registered the same yield levels (3029–3375 kg/ha). Fertigation at the rate of 75% NPK increased the yield of cocoa by 52% over VCE alone. The results indicated that drip fertigation increases the productivity, but precision application of N and K is required for sustaining the yields [49].

1.5   CONSTRAINTS/LIMITATION OF FERTIGATION

- **Contamination of drinking water:** Generally, irrigation water forms part of the drinking water network in many farming system. As these fertilizers are toxic, suitable non-return valves should be placed to avoid the fertilizers mixing into the water source. Another option is routing water supply through a separate tank and not from the water source itself (open well).
- **Corrosion:** The metallic parts of the equipment are highly prone to corrosion. Sensitive parts of the equipment must be made of protected or resistant materials and extra care should be taken, while filling the tanks.
- **Fertilizer suitability:** The method is suitable for soluble fertilizers. However, some fertilizers such as superphosphate and calcium ammonium phosphate are having low solubility, hence are not suitable, and may clog the pipes/emitters.
- **Availability of material:** In India, the required soluble fertilizer and grades are not available freely and the cost is also high.
- **Application of organic manure:** It is difficult to apply the conventional organic manures through drip fertigation. Farmers need to prepare the slurry of manure or vermiwash and it needs to be filtered properly before mixing into the system. Besides, liquid organic fertilizers are not available easily in India; the cost is also high.

1.6 POLICY AND INSTITUTIONAL REASONS

- In India, there is a subsidy policy for normal NPK fertilizers in specified grades. However, for fertigation, the requirement of fertilizer is in different grades and it should be 100% water soluble for its effective application. The fertigation material is either not available in desired form or available at higher price, than the conventional fertilizers. The Government should adopt a fertilizer policy in such a way that the manufacturers of fully soluble fertilizer are not in disadvantage as compared to conventional fertilizer manufacturers.
- Once the fertigation practice is being followed along with drip irrigation, system may cause higher clogging, localized acidity in soils etc. The farmers must be trained to adopt fertigation in a sustainable way.
- There is lack of research and developmental information in respect of its rate of application, amount applied, and frequency adopted. However, research efforts are being focused on this aspect but there is a lack of information in respect of varied agro-climatic conditions and crops.

1.7 POLICY RECOMMENDATIONS AND SUGGESTIONS

- Reducing the capital cost and increasing technical knowledge of farmers will help to improve drip fertigation adoption.
- Fertigation (application of fertilizers through drip irrigation) is not practiced by most of the farmers even though they adopt drip irrigation system. Hence, it should be made mandatory under government subsidy schemes, that fertigation should be part of drip irrigation programs to increase the crop productivity and profitability of the farmers.
- A special purpose vehicle (SPV)/project cell should be created for implementation and follow up of drip fertigation. In each district, a separate project cell is to be created with Project Director, Drip Engineers, and other support staff for drip irrigation maintenance which some of the Indian states like Andhra Pradesh, Gujarat etc. are already implementing.
- Water-soluble fertilizers (both organic and chemical fertilizers) need to be included under fertilizer subsidy policy.
- One of the major recommendations is the need of technical support. Capacity building of the implementing team is necessary, which, in turn, can train the farmers on the use of drip fertigation.

1.8 CONCLUSIONS

Fertigation is the process by which, both water and fertilizers are delivered to the crop simultaneously through a drip irrigation system. Fertigation ensures that essential nutrients are supplied precisely at the area of most intensive root activity according to the specific requirements of crops and type of soil, resulting in higher crop yield and quality of the produce. Fertigation was first developed for field and horticultural crops, and later used on tree plantations. Nowadays, with the advent of hi-tech agriculture in India (especially Kerala and Tamil Nadu) viz., polyhouses, precision farming, and greenhouse cultivation, the use of fertigation with automatic scheduling of irrigation cycle for agriculture purpose is gaining importance in these parts of India.

According to several authors, drip fertigation will continue to expand and slowly replace traditional flood irrigation wherever population demand for fresh water will put pressure on water resources. This holds good for semi-arid and humid tropical regions of India since the pressure on water resources are increasing day by day. Labor costs are also an important factor in the transformation from flood or canal irrigation to permanent fertigation systems. As farmers have already shifted from subsistence agriculture to profitable agriculture, the shift to fertigation is inevitable.

Vegetables have been found particularly responsive to fertigation due to their spacing nature, continuous need of water and nutrients at optimal rate to give high yield with good quality, high capital turnover to investments, and maybe their cultivation by more skilled and progressive farmers. Even though the initial cost of establishing the fertigation system is higher, in long-term basis, it is economical compared to conventional methods of fertilization as it brings down the cost of cultivation. However, to get the desired results, it requires higher management skills at operator level like selection of fertilizers, timing, and rate of fertilizer injection, watering schedule, as well as the maintenance of the system. Users may face some practical problems like clogging of emitters, pest and disease problems, malfunctioning of the drip fertigation system etc. But such problems can be overcome through effective management skills of the users which build up over the time with the use of the system and technical knowledge of the system. Therefore, to make the agriculture sustainable and economically viable and to ensure food and nutritional security of the burgeoning population, there is need to promote the drip fertigation at large scale by the concerned stakeholders. There is need to develop recommendations for the most suitable fertilizer formulations including the basic nutrients (NPK) and microelements according to the local soil type, climate, crops, and their physiological stages, and other factors like nutrient mobility in the soil and salinity. Further, there is need to work on reducing the initial cost of establishment through continuous research and development in technology which suits best to Indian conditions. Authors feel that the basic topics on the combined use of plant nutrients with irrigation as discussed in this chapter will be of benefit to farmers, researchers, and all stakeholders all over the world for the efficient use of water and nutrients in agricultural production systems.

ACKNOWLEDGMENTS

The authors are thankful to the Executive Director of CWRDM (Center for Water Resources Development and Management) for providing the necessary support and encouragement. Data collected from various projects with funding support from TIFAC, SERB, and SSTP, Department of Science and Technology, Government of India and R&D Center, EID Parry is gratefully acknowledged.

KEYWORDS

- **constraints**
- **drip irrigation**
- **drip irrigation adoption**
- **fertigation**
- **GOI**
- **India**
- **water requirement**
- **water saving**
- **WUE**

REFERENCES

1. Alva, A. K.; Paramasivam, S.; Graham, W. D.; Wheaton, T. A. Best Nitrogen and Irrigation Management Practices for Citrus Production in Sandy Soils. *Water. Air. Soil. Pollut.* **2003,** *143,* 139–154.

2. Bar-Yosef, B. Advances in Fertigation. *Adv. Agron.* **1999,** *65,* 1–70.

3. Bhakare, B. D.; Fatkal, Y. D. Influence of Micro Irrigation and Fertilizer Levels through Fertigation on Growth, Yield and Quality of Onion Seed. *J. Water Manage.* **2008,** *16*(1), 35–39.

4. Bhanurekha, K.; Govind Reddy, M.; Mahavishnan, K. Nitrogen and Water Use Efficiency of Okra (*Abelmoschus Esculentus* L. Moench) as Influenced by Drip Fertigation. *J. Trop. Agric.* **2005,** *43,* 45–49.

5. Darwish, T.; Atallah, T.; Hajhasan, S.; Chranek, A. Management of Nitrogen by Fertigation of Potato in Lebanon. *Nutr. Cycl. Agroecosys.* 2003, *67*(1), 1–11.

6. Dash, N. K. Drip Irrigation and Micro-Irrigation Potential and Prosperity for Sugarcane Cultivation with Reference to Orissa. Vasantdada Sugar Institute (VSI): Manjari (BK), Pune, Maharashtra, India, 1998; pp III7– III12.

7. GOI. *Report of Task Force on Micro Irrigation.* Ministry of Agriculture, Govt. of India: New Delhi, India, January, 2004.

8. GOI. *Report of the Task Force on Irrigation.* Planning commission, Govt. of India: New Delhi, India, 2009.

9. GOTN. *Study on Precision Farming Project by Monitoring and Evaluation Unit.* Directorate of Agriculture, Government of Tamil Nadu - Report 221, Tamil Nadu, India, 2007.

10. Goyal, Megh R. *Research Advances in Sustainable Micro Irrigation.* Apple Academic Press Inc.: Oakville, ON, Canada, 2015; Vol. 1–10.

11. Goyal, Megh R. *Innovations and Challenges in Micro Irrigation.* Apple Academic Press Inc.: Oakville, ON, Canada, 2016; Vol. 1–4.

12. Hanson, B. R.; Simunek, J.; Hopmans, J. W. Evaluation of Urea–Ammonium–Nitrate Fertigation with Drip Irrigation using Numerical Modeling. *Agr. Water Manage.* **2006,** *86,* 102–113.

13. ICID. *Annual Report of International Commission on Irrigation and Drainage.* ICID: New Delhi, India, 2013–14.

14. INCID. *Drip Irrigation in India.* Indian National Committee on Irrigation and Drainage: New Delhi, India, 1994.

15. Jat, R.; Wani, S.; Sahrawat, K.; Singh, P.; Dhaka, B. L. Fertigation in Vegetable Crops for Higher Productivity and Resource Use Efficiency. *Ind. J. Fert.* **2011,** *7*(3), 22–37.

16. Jayakumar, M.; Surendran, U.; Manickasundaram, P. Drip Fertigation Effects on Yield, Nutrient Uptake and Soil Fertility of Cotton in Semiarid Tropics. *Int. J. Plant Prod.* **2014,** *8*(3), 375–390.

17. Jayakumar, M.; Surendran, U.; Manickasundaram, P. Drip Fertigation Program on Growth, Crop Productivity, Water, and Fertilizer-Use Efficiency of Cotton in Semiarid Tropical Region of India. *Commun. Soil Sci. Plant Anal.* **2015,** *46*(3), 293–304.

18. Kafkafi, U. Global Aspects of Fertigation Usage. In *Fertigation: Optimizing the Utilization of Water and Nutrients,* Fertigation Proceedings: Selected Papers of the IPI-NATESC-CAU-CAAS International Symposium on Fertigation, Beijing, PR China, September 20–24, 2005; Imas, P.; Price, R.; Eds.; China Agriculture Press: Beijing, PR China, 2005.

19. Kumar, M. D.; Singh, K.; Singh, O. P.; Shiyani, R. L. *Impact of Water Saving and Energy Saving Irrigation Technologies in Gujarat. Research Report 2.* Indian Natural and Economic Resource Management: Anand, Gujarat, India, 2006.

20. Kumar, M. D. *Dripping Water to a Water Guzzler: Techno-Economic Evaluation of Efficiency of Drip Irrigation in Alfalfa.* Paper presented at 2nd International Conference

of the Asia Pacific Association of Hydrology and Water Resources (APHW), Singapore, July 5–9, 2004.

21. Kumar, M. D. Market Instruments for Demand Management in the Face of Scarcity and Overuse of Water in Gujarat, India. *Water Policy.* **2001,** *5*(3), 387–403.

22. Vijayakumar G.; Tamilmani, D.; Selvaraj, P. K. Irrigation and Fertigation Scheduling under Drip Irrigation in Brinjal (*Solanum melongena* L.) Crop. *IJBSM.* **2010,** *1*(2), 72–76.

23. Mahajan, G.; Singh, K. G. Response of Greenhouse Tomato to Irrigation and Fertigation. *Agr. Water Manage.* **2006,** *8*(4), 202–206.

24. Mohammad, M. J. Utilization of Applied Fertilizer Nitrogen and Irrigation Water by Drip-Fertigated Squash as Determined by Nuclear and Traditional Techniques. *Nutr. Cycl. Agroecosys.* **2004,** *68*(1), 1–11.

25. Mohammad, M. J. Squash Yield, Nutrient Content and Soil Fertility Parameters in Response to Fertilizer Application and Rates of Nitrogen Fertigation. *Nutr. Cycl. Agroecosys.* **2004,** *68*(2), 99–108.

26. MOWR. *Report of the Working Group on Water Availability for Use.* National Commission for Integrated Water Resources Development Plan Ministry of Water Resources, Government of India: New Delhi, India, 1999.

27. Muralikrishnasamy, S.; Veerabadran, V.; Krishnasamy, S.; Kumar, V.; Sakthivel, S. *Micro Sprinkler Irrigation and Fertigation in Onion (Allium cepa).* Paper presented at the 7th International Micro Irrigation Congress, September 10–16, Kuala Lumpur, Malaysia, 2006.

28. Namara, R. E.; Upadhyay, B.; Nagar, R. K. *Adoption and Impacts of Microirrigation Technologies: Empirical Results from Selected Localities of Maharashtra and Gujarat States of India. Research Report 93.* International Water Management Institute: Colombo, Sri Lanka, 2005.

29. Narayanamoorthy, A. Impact of Drip Irrigation on Consumption of Water and Electricity. *Asian Econ. Rev.* **1996,** *38*(3), 350–364.

30. Narayanamoorthy, A. Economic Viability of Drip Irrigation: An Empirical Analysis from Maharashtra. *Ind. J. Agr. Econ.* **1997,** *52*(4), 728–739.

31. Narayanamoorthy, A. *Impact of Drip Irrigation on Sugarcane Cultivation in Maharashtra.* Agro-Economic Research Center, Gokhale Institute of Politics and Economics: Pune, India, June, 2001.

32. Narayanamoorthy, A. *Efficiency of Irrigation: A Case of Drip Irrigation.* Occasional Paper 45. Department of Economic Analysis and Research, National Bank for Agriculture and Rural Development: Mumbai, India, 2005.

33. Narayanamoorthy, A. Impact Assessment of Drip Irrigation in India: The Case of Sugarcane. *Dev. Policy Rev.* **2008,** *22*(4), 443–462.

34. Neilsen, G. H.; Neilsen, D.; Peryea, F. J. Response of Soil and Irrigated Fruit Trees to Fertigation or Broadcast Applications of Nitrogen, Phosphorus and Potassium. *Hortic. Technol.* **1999,** *9,* 393–401.

35. Neilsen, D.; Neilsen, G. H. Efficient Use of Nitrogen and Water in High Density Apple Orchards. *Hortic. Technol.* **2002,** *12,* 19–25.

36. Neilsen, G. H.; Neilsen, D.; Herbert, L. C.; Hogue, E. J. Response of Apple to Fertigation of N and K under Conditions Susceptible to the Development of K Deficiency. *J. Am. Soc. Hortic. Sci.* **2004,** *129,* 26–31.

37. Ouyang, D. S.; Mackenzie, A. F.; Fan, M. X. Availability of Banded Triple Super Phosphate with Urea and Phosphorus Use Efficiency by Corn. *Nutr. Cycl. Agr. Ecosys.* **1999,** *53,* 237–247.

38. Patel, N.; Rajput, T. P. S. Simulation and Modelling of Water Movement in Potato (Solanum tuberosum). *Ind. J. Agr. Sci.* **2011,** *81*(1), 15–25.

39. Pedersen, S.; Abrahamsen, P.; Plauborg, F. Economic Feasibility of Drip and Fertigation Systems in Europe. In: Proc: NJF Congress: Trends and Perspectives in Agriculture, Copenhagen, Denmark Drip and fertigation system, by Nordic Association of Agricultural Scientists: Uppsala, Sweden, 2007.

40. Howlett, P.; Velkar, A. Agri-Technologies and Travelling Facts: A Case Study of Extension Education in Tamil Nadu, India. Working Paper No. 35/08. London School of Economics and Political Science: London, UK, November, 2008; pp14.

41. Rane, N. B. *Development, Scope and Future Potential of Fertigation in India,* Proceedings of National Seminar on Advances in Micro Irrigation, NCPAH, Ministry of Agriculture, GOI, New Delhi, India, Feb 15–16, 2011; pp 44.

42. Ravi, B.; Sujatha, S.; Balasimha, D. Impact of Drip Fertigation on Productivity of Arecanut (Areca catechu L.). *Agr. Water Manage.* **2007,** *90*(1), 101–111.

43. Shedeed, I.; Sahar, M.; Zaghloul, A.; Yassen, A. Effect of Method and Rate of Fertilizer Application under Drip Irrigation on Yield and Nutrient Uptake by Tomato. *Ozean J. Appl. Sci.* **2009,** *2*(2), 139–147.

44. Shirgure, P. S.; Ram, L.; Marathe, R. A.; Yadav, R. P. Effect of Nitrogen Fertigation on Vegetative Growth and Leaf Nutrient Content of Acid Lime (Citrus aurantifolia, Swingle) in Central India. *Ind. J. Soil Conserv.* **1999,** *27*(1), 45–49.

45. Sivanappan, R. K. *Irrigation Water Management for Sugarcane.* National Seminar on Irrigation Water Management for Sugarcane. Vasantdada Sugar Institute: Manjari (BK), Punc, Maharashtra, India, 1998.

46. Sivanappan, R. K. Prospects of Micro-Irrigation in India. *Irrigat. Drain. Syst.* **1994,** *8*(1), 49–58.

47. Sivanappan, R. K. *Strengths and Weaknesses of Growth of Drip Irrigation in India,* Proceedings of the GOI Short-Term Training on Micro Irrigation for Sustainable Agriculture, June 19–21, 2002; WTC, Tamil Nadu Agricultural University: Coimbatore, India, 2002.

48. Sne, M. *Micro Irrigation in Arid and Semi-Arid Regions: Guidelines for planning and Design;* Kulkarni, S.A., Ed.; International Commission on Irrigation and Drainage (ICID-CIID): New Delhi, India, 2006.

49. Sujatha, S.; Ravi, B. Impact of Drip Fertigation on Arecanut–Cocoa System in Humid Tropics of India. *Agroforest. Syst.* 2013, *87*(3), 643–656.

50. Surendran, U.; Ramesh, M.; Vani, D. *An Analysis on the Impact of Drip Irrigation on Productivity of Sugarcane.* EID Parry (I) Ltd.: Chennai, India, 2007; Publication 37.

51. Surendran, U.; Sushanth, C. M.; George Mammen; Joseph, E. J. Modeling the Impacts of Increase in Temperature on Irrigation Water Requirements in Palakkad District – A Case Study in Humid Tropical Kerala. *J. Water Clim. Change.* **2014,** *5*(3), 471–487.

52. Trickle Irrigation Manual. Saskatchewan Trickle Irrigation Manual, 2011. http://www.irrigationsaskatchewan.com

53. Veerana, H. K.; Khalak, A.; Farooqui, A. A.; Sujith, G. M. Effect of Fertigation with Normal and Water-Soluble Fertilizers Compared to Drip and Furrow Methods on Yield Fertilizer and Irrigation Water Use Efficiency in Chilli. Central Board of Irrigation and Power (CBIP): New Delhi, India, 2001; Publication 282, pp 461.

54. Verma, S.; Tsephal, S.; Jose, T. Pepsee Systems: Grass Root Innovation under Groundwater Stress. *Water Policy.* **2004,** *6,* 1–16.

CHAPTER 2

DESIGN OF TRICKLE IRRIGATION SYSTEMS

I. PAI WU and HARRIS M. GITLIN

Department of Molecular Biosciences and Bioengineering, University of Hawaii at Macoa, Honolulu, HI, USA

CONTENTS

ABSTRACT

This chapter discusses basic concepts of hydraulics needed to develop the design charts for the lateral and secondary lines of a drip irrigation system. It provides a thorough explanation of the chart and design procedures for lateral and secondary lines, on uniform and non-uniform slopes. It also explains the design procedure for lateral and secondary lines of different sizes and the design of the main line. These procedures are carefully outlined. To enhance the understanding of the design procedure for each case, the design examples are presented.

2.1   INTRODUCTION

A drip irrigation system consists of main lines, secondary lines, and lateral lines [1]. There are other important components such as filters, regulators of pressure, indicators of pressure, valves, injectors of fertilizer, and so forth (Fig. 2.1). The lateral lines can be polyethylene tubes combined with drippers, or simply of low pressure plastic pipe with orifices. These are designed to distribute water to the field with an acceptable degree of uniformity [8–10]. The secondary line acts as a control system, which can adjust the pressure of water to supply the quantity of flow required in each lateral line. Also it is used to control the time of irrigation in individual field. The main line serves as a system of transportation to supply the total quantity of water required in the irrigation system. This chapter is a combined version of "*Drip irrigation systems design*" bulletins # 144 and 156 by the Cooperative Agricultural Extension Service at the University of Hawaii [11].

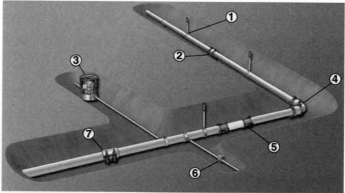

FIGURE 2.1   Components a trickle irrigation system: 1. Main line, 6. Submain, 3. Pump.

2.2 BASIC CONCEPTS OF HYDRAULICS

Plastic pipes of different sizes can be considered like smooth conduits. The Blasius formula can be used to determine the turbulent flow in a smooth conduit [5, 6]. The empirical equation of Williams and Hazen [3] for $C = 150$ is given in Table 2.1. The Eq 2.1 is used to determine the loss or fall in energy for the section of main line. In Eq 2.1 and in Table 2.1:

ΔH = The energy loss by friction in feet or meters.
ΔL = Length of the pipe section in feet or meters.
Q = Total discharge in the pipe in gallons per minute (gpm) or liters per second (lps).
D = Internal diameter of the pipe in inches or centimeters.

The condition of the flow in the secondary or lateral line is constant and varies especially with the flows of the lateral line. Since the discharge in the line decreases with the length, the loss will be less than is given by Eq 2.1. The loss of energy due to friction for the lateral or secondary lines is shown in Eq 2.2. In Eq 2.2 and in Table 2.1:

ΔH = Total energy loss by friction at end of the lateral line (or secondary) in feet or meters.
Q = Total discharge at the entrance of the lateral line (or secondary) in gpm or lps.
D = Internal diameter of the lateral line (or secondary) in inches or centimeters.
L = Total length of the lateral line (or secondary) in feet or meters.

TABLE 2.1 Equations for Basic Concepts of the Hydraulics.

$\Delta H = 9.76 \times 10^{-4} \times (Q^{1.852}/D^{4.871}) \times \Delta L$	FPS	(2.1a)
$\Delta H = 15.27 \times (Q^{1.852}/D^{4.871}) \times \Delta L$	MKS	(2.1b)
$\Delta H = 3.42 \times 10^{-4} \times (Q^{1.852}/D^{4.871}) \times L$	FPS	(2.2a)
$\Delta H = 5.35 \times (Q^{1.852}/D^{4.871}) \times L$	MKS	(2.2b)
$R^i = 1 - [1 - i]^{2.852}$	–	(2.3)
$(dh/dl) = -S_f \pm S_o$	–	(2.4)
$q = C \times (\sqrt{h})$ o $q = C \times (h)^{0.5}$	–	(2.5)
$q_{var} = 1 - [1 - h_{var}]^{0.5}$	–	(2.6)
$q_{var} = [q_{max} - q_{min}]/q_{max}$	–	(2.7)
$h_{var} = [h_{max} - h]/h_{max}$	–	(2.8)

When the discharge in the lateral line decreases with the length, the gradient line of energy is an exponential curve [2, 5] instead of a straight line. The form of the gradient line can be expressed as the gradient line of energy without dimensions as shown in Eq 2.3. In Eq 2.3 and in Table 2.1:

R_i = $\Delta H_i/\Delta H$, and it is known as energy drop ratio.
ΔH = Total loss of energy determined by Eq 2.2.
ΔH_i = Total loss of energy expressed in feet or meters, for length ratio of i ($i = 1/L$).
L = Total length in feet or meters.
L = Given length, measured since the final section of the line in feet or meters.
i = $1/L$.

Eq 2.3 can be used to determine the pattern of energy loss along a lateral line (or secondary), when the total loss in energy is known. The variation in pressure (the change in pressure along the length) can be determined like a lineal combination of slope of the energy and the slope of the line, assuming that the change of velocity head is small or insignificant. This is shown in Eq 2.4. This relation can be used for non-uniform and uniform slopes [6]. In Eq 2.4 and in Table 2.1:

S_f = Energy slope or the energy gradient line.
$\pm S_o$ – Slope of the line, with positive sign when the line is down the slope, and with negative sign when the line is above the slope.

Hydraulically, the variation in pressure along a lateral line will cause a variation in the flow of the dripper along the lateral line. A variation in pressure along the secondary line would cause a variation in the flow of the dripper along the lateral line (toward each lateral line) and along the secondary line. For most common dripper types, and assuming a turbulent flow in the lateral line, the discharge of the dripper (or the lateral lines flow for the secondary one) and of the pressure head can be expressed as a simple function by Eq 2.5. In Table 2.1 and in Eq 2.5:

q = Flow of dripper (or flow toward the lateral line).
h = Pressure head.

The Eqs 2.6 and 2.7 describe the relation between the variation in pressure and the variation in the dripper flow. The variation in pressure is given by the Eq 2.8. In Eqs 2.6, 2.7, and 2.8 and Table 2.1:

q_{var} = Dripper flow variation for lateral line (or secondary line).

q_{max} = Maximum dripper flow for lateral line (or secondary line).

q_{min} = Minimum dripper flow for the lateral line (or secondary line).

h_{max} = Maximum pressure head along the lateral line (or secondary line).

h_{min} = Minimum pressure head along the lateral line (or secondary line).

h_{var} = Variation in pressure head along the lateral line (or secondary line).

The design criteria [6] is a variation in the dripper flow of less than the 20% (approximately 40% for variation in pressure) for the design of lateral line and a variation in the flow of the lateral line less than 5% (approximately 10% for variation in pressure) for the design of the secondary line.

2.3 CHARTS AND DESIGN PROCEDURES

The design charts were based on the hydraulics of the drip lines (Section 2.2 in this chapter) and were developed with a computer simulation [4, 6, 7] for the design of lateral line, secondary line, and main line. The design charts and its procedures are described in the following section:

2.3.1 DESIGN CHARTS FOR THE LATERAL LINE ON UNIFORM SLOPES

The design charts for lateral line of 0.5 inch were developed as shown in Figures 2.2 to 2.5. The Figures 2.6 and 2.7 show design charts for lateral line of 12 and 16 mm, respectively. The procedure is:

Step 1: Establish along one of the lateral lines: the length of lateral line (L), the operational pressure head (H), the ratio L/H, and the total discharge Q in gpm or lps.

Step 2: Move vertically from L/H in the third quadrant to the total given discharge (gpm or lps) in the second quadrant; then establish a horizontal line toward the first quadrant.

Step 3: Move horizontally from L/H (third quadrant) to the percent slope in the fourth quadrant; then establish a vertical line toward the first quadrant.

Step 4: The point of intersection of these two lines in the first quadrant determines the acceptability of the design.

Desirable: A coefficient of uniformity (CU) over 98%; equivalent to a variation in the dripper flow less than 10% or a variation in pressure less than 20%.

Acceptable: CU between 95 and 98%; equivalent to a variation in the dripper flow around 10–20% or variation in pressure of 20–40%.

Not acceptable: CU less than 95%; equivalent to a variation in the dripper flow over 20% or a variation in pressure greater than 40%.

2.3.2 GENERAL DESIGN CHARTS FOR THE LATERAL AND SECONDARY LINES ON UNIFORM SLOPES

The design charts were developed for the lateral lines (Fig. 2.2 to 2.5 for the English system, and Fig. 2.6 and 2.7 for metric system). The Figures 2.8 and 2.9 are dimensionless charts for all sizes of the line; and these can be used for the secondary and lateral lines. The Figure 2.10 is for the English system and the Figure 2.11 is for the metric system. These figures are used to determine H/L from the total discharge and the size of the line.

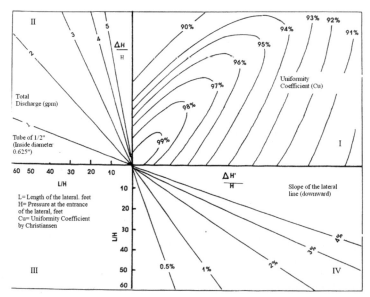

FIGURE 2.2   Design chart for the lateral line of ½ inch (downward of the slope).

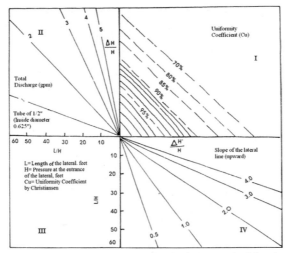

FIGURE 2.3   Design chart of the lateral line of ½ inch (upward of the slope).

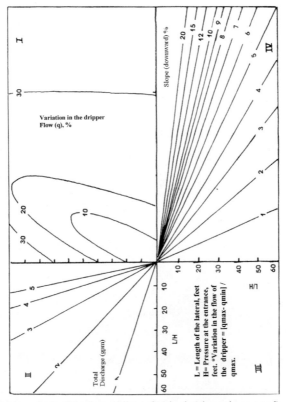

FIGURE 2.4   Design chart of the lateral line of ½ inch (slope downward).

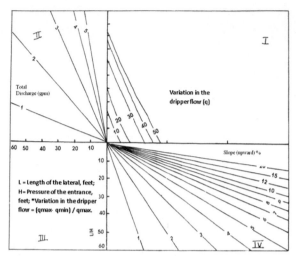

FIGURE 2.5   Design charts of the lateral line of ½ inch (upward of the slope).

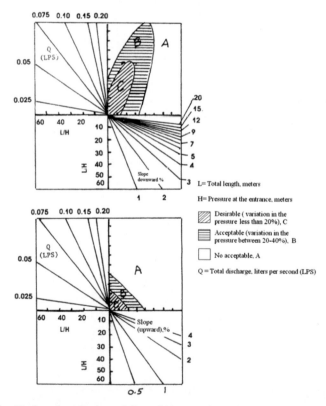

FIGURE 2.6   Design chart for lateral line of 12 mm (slope downward and upward).

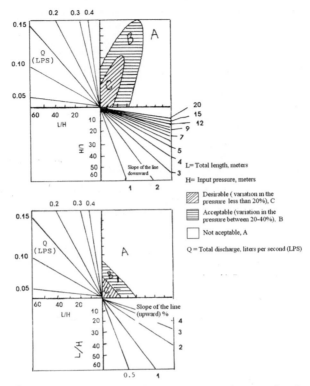

FIGURE 2.7 Design chart for lateral line of 16 mm (slope downward and upward).

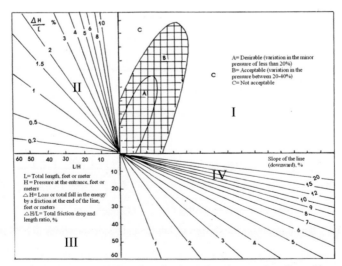

FIGURE 2.8 Dimensionless design chart for lateral and secondary lines (downward of the slope).

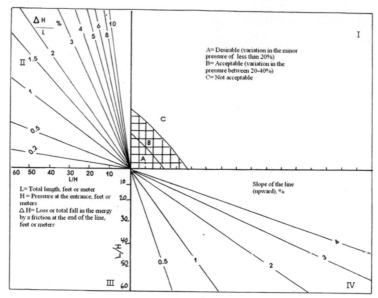

FIGURE 2.9   Dimensionless design chart for the lateral and secondary lines (upward of the slope).

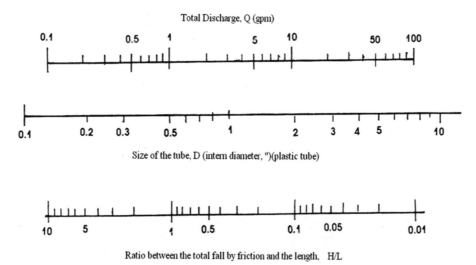

FIGURE 2.10   Nomograph for the design of lateral and secondary lines in FPS units.

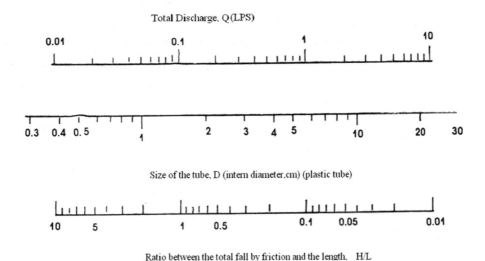

FIGURE 2.11 Nomograph for the design of lateral and secondary lines in metric units.

These design charts can be used to revise the acceptability of the design, if the size of the lateral line is given, or for selecting an appropriate size of lateral lines to comply with the design criteria. The design procedure is:

A. To revise the acceptability of the design (if we know the size of the lateral or secondary lines):

Step 1: Establish a trial L/H and the total discharge.

Step 2: One should use the total discharge and the size of the lateral line to determine $\Delta H/L$ (Fig. 2.10 or 2.11).

Step 3: Move vertically from L/H (third quadrant) toward the specific value of $\Delta H/L$ in the second quadrant (Fig. 2.8 and 2.9); then establish a horizontal line toward the first quadrant.

Step 4: Move horizontally from L/H toward the slope of the line in the fourth quadrant; then establish a vertical line toward the first quadrant.

Step 5: The point of intersection of these two lines in the first quadrant will determine the acceptability of the design:

Zone A: Desirable, variation in the dripper flow of less than 10%.
Zone B: Acceptable, variation in the dripper flow from 10 to 20%.
Zone C: Not acceptable, variation in the dripper flow of greater than 20%.

B. To select the appropriate size of the lateral or secondary line.

Step 1: Establish a trial value of L/H and the total discharge.

Step 2: Move horizontally from L/H toward the slope of the line (up or downward) in the fourth quadrant. From this point, establish a vertical line toward the first quadrant.

Step 3: Establish a point along this line in the first quadrant in the upper margin of the acceptable region depending on the design criteria. From this point, establish a horizontal line toward the second quadrant.

Step 4: Establish a vertical line toward the second quadrant for the value L/H in such a way to intercept the point in the horizontal line indicated in the Step 3.

Step 5: Determine the value of $\Delta H/L$ in the second quadrant for this point.

Step 6: From Figures 2.10 and 2.11, the total discharge is calculated. The values of $\Delta H/L$ and the minimum size of the lateral lines are established in agreement with the design criteria.

2.3.3   DESIGN OF LATERAL LINE ON NON-UNIFORM SLOPES

A simple design chart for lateral line on non-uniform slopes [9] is shown in Figure 2.12. The chart is dimensionless, so that it can be used for English and metric units. The design procedure is:

Step 1: Divide the profile of the non-uniform slope in several sections, so that each section can be considered as a uniform slope. Determine the slope of each section; calculate the gain (or the loss) of energy in each section due to its slope, and find the total energy gain by the slopes of any section along the line ($\Delta H'_i$).

Step 2: Plot the pattern of the non-uniform slope, using dimensionless design chart in Figure 2.12: $1/L$ vs. $\Delta H'_i/L$ in the first quadrant.

Step 3: Determine the total energy loss by friction (ΔH), using Eq 2.2. Calculate the ratio of total loss in energy (ΔH) and the operational pressure head, $\Delta H/H$. One may also use Figure 2.10 or 2.11.

Step 4: Determine the ratio of the length of the lateral lines (L) to the operational pressure head (H): L/H.

Step 5: Select any point on the profile of the non-uniform slope in the first quadrant: usually a point between two slopes or the middle of the section.

Step 6: From this point, draw a vertical line downward of the specific value of L/H in the fourth quadrant, and establish a horizontal line toward the third quadrant.

Step 7: Also from the point as indicated in Step 5, draw a horizontal line at specific $\Delta H/H$ in the second quadrant, then establish a line toward the third quadrant.

Step 8: Locate the point of intersection of these two lines in the third quadrant. It will give the variation in the pressure.

Step 9: Repeat the same procedure for various other points in the profile of the non-uniform slope (Step 1). Revise the variations in pressure for these points of the operational pressure head.

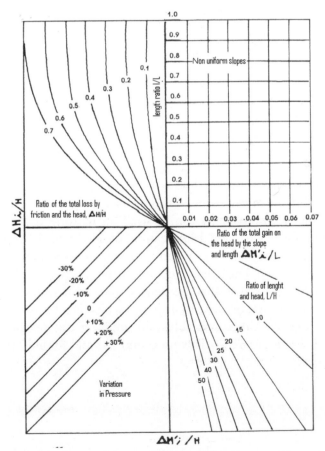

FIGURE 2.12 Dimensionless design chart for lateral line for non-uniform slopes.

2.3.4 DESIGN OF TRICKLE IRRIGATION SYSTEM FOR DIFFERENT SIZES OF LATERAL AND SECONDARY LINES [10]

The most of the lateral and secondary lines are designed for a pipe of selected size. The energy gradient line for a lateral line of a simple size is an exponential curve, which is used as a basis to design lateral or secondary line in drip irrigation for uniform slopes and non-uniform slopes. However, the length of one of the lateral and secondary lines can be relatively large under specific conditions and for non-uniform slopes. The design of the lateral and secondary lines is a series of pipes of different sizes.

If the lateral or secondary line of individual size is designed so that the total energy loss by friction (ΔH) is balanced by the energy gain (H) at the end of the line, then the maximum variation in pressure head will be ($0.36 \Delta H_i$) or ($0.36 S_o L$), where:

S_o = Slope of the lateral or secondary lines.

L = Length of the line.

This is caused by the shape of curve of the energy gradient line. The maximum variation in pressure will occur near the middle of the individual section of the line. When a series of different sizes in the design of lateral and secondary lines can be used, the maximum variation in pressure head can be reduced. If one can use three, four, or more different sizes, the variation in pressure head along the line will be less, as shown in Figure 2.13. The slope of the line of each section can be used as the energy to design the size of the line. This procedure can be used for uniform and non-uniform slopes. If a secondary or lateral line can be divided into different sections and different sizes for each section can be designed, then it has been shown that the energy gradient line of each section can be approximated as a straight line. It has also been shown that the discharge of each section can be used to calculate the total energy loss by friction, which is a basis for the design of size of tubing. With these variables, the engineer can design sections of lateral lines using Figure 2.14. This procedure is valid for uniform and non-uniform slopes.

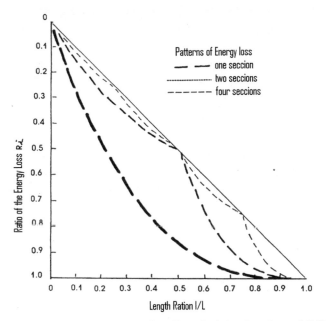

FIGURE 2.13   Dimensionless energy gradient lines for irrigation pipes of different sizes.

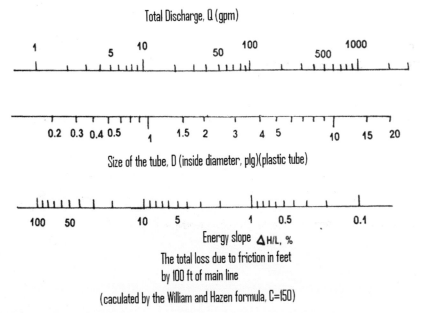

FIGURE 2.14   Nomograph for the design of main and secondary lines in FPS units (for multiple sections with varying sizes).

2.3.5 SIMPLIFIED DESIGN CHART FOR MAIN AND SECONDARY LINES

Design charts similar to Figures 2.2 and 2.3 for English system and Figures 2.6 and 2.7 for metric system can be developed for the design of secondary lines. The design chart for non-uniform slopes can be utilized for the design of secondary lines. However, for shorter length of the secondary line, one can assume a uniform slope. Two design charts for secondary lines are developed: one for lines with slopes ≥ 0.5%, and another for secondary lines with slopes < 0.5%. The design charts are given in Figures 2.15 and 2.16 for English units and Figures 2.17 and 2.18 for metric units. Generally, the length of the secondary line is short, from 66 to 200 ft (= 20–60 m). If the operational pressure head is from 20 to 30 ft (= 6–9 m), the two simplified design charts will give a variation in flow of less than 5% for lateral line. The design procedure is:

Step 1: Determine the total discharge Q, for the secondary line.

Step 2: Determine the ratio of the length of the line to the pressure head: L/H.

Step 3: Determine the slope of the secondary line. If the slope is equal or greater than 0.5%, use Figure 2.14 or 2.16 to design the size of the secondary line.

Step 4: If the slope is less than 0.5%, use Figure 2.16 or 2.18 to determine the size of the secondary line.

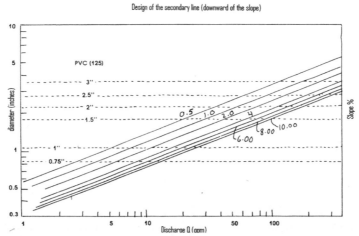

FIGURE 2.15 Design chart for the secondary line for slopes greater than 0.5% (FPS units).

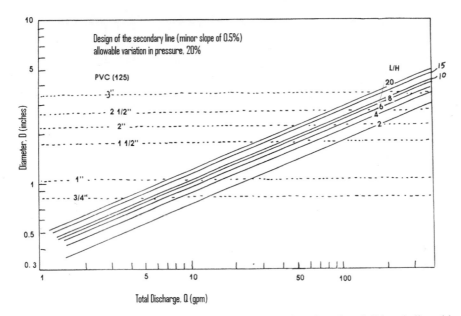

FIGURE 2.16 Design chart for the secondary line for slope less than 0.5% and allowable variation of pressure of 20% in FPS units.

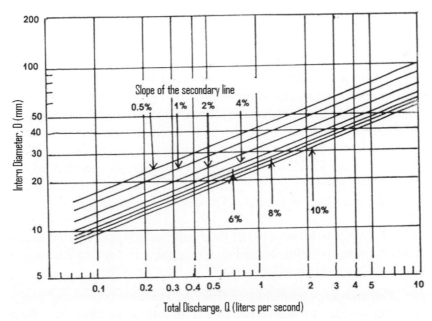

FIGURE 2.17 Design chart for the secondary line for slopes equal or greater than 0.5% (MKS units).

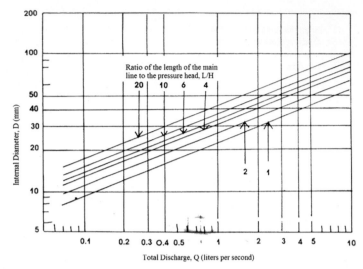

FIGURE 2.18   Design chart for the secondary line for slope less than 0.5% and allowable variation in pressure of 10% (MKS units).

2.3.6   DESIGN OF MAIN LINE

The design of the main line is not a problem. The main line is designed based on total discharge, the length, and the allowable energy loss according to Eq 2.1 in Table 2.1. When the main line supplies water to many secondary lines (or to many subfields), the total discharge in each section of main line diminishes along the length of the main line. The size of the main line for different sections will depend on the shape of the energy gradient line for the main line. The optimum solution can be obtained using a computer simulation. The concept of straight energy gradient line was developed [4] for the design of the main line, which simplifies the design procedure. Given that the design procedure is simple, it can be used to design alternate arrangements of the main lines in the field. The design procedure is:

Step 1: Plot the profiles (slopes) of the main line and the required pressure head for the drip irrigation according to Figure 2.19 for English units and Figure 2.20 for metric units.

Step 2: Plot an energy gradient line from the available operational pressure heads to the profile of required pressure (Fig. 2.19 and 2.20) so that for any point along the main line, the gradient of energy line agrees with the profile of required pressure.

Step 3: Determine the slope of the straight energy gradient line: $\Delta H/L$.

Step 4: Determine the discharge required for each section of the main line.

Step 5: Design the size of the main line using Figure 2.21 or 2.22 based on the slope of energy (determined in the Step 3) and the total discharge (determined in the Step 4) for each section of the main line.

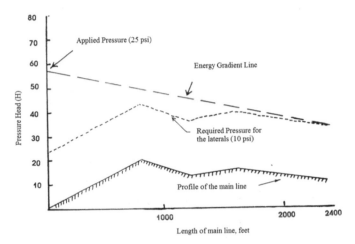

FIGURE 2.19 Profile of energy gradient line and profile of the main line (FPS units), pressure head in feet.

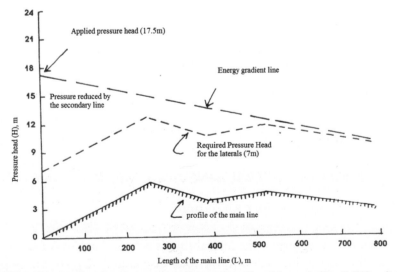

FIGURE 2.20 Profile of energy gradient line and profile of the main line (MKS units).

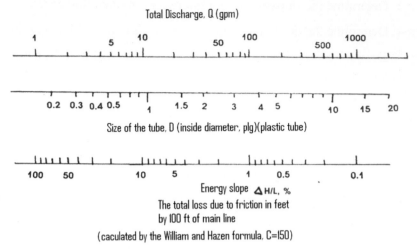

FIGURE 2.21   Nomograph for the design of main and secondary lines (FPS units).

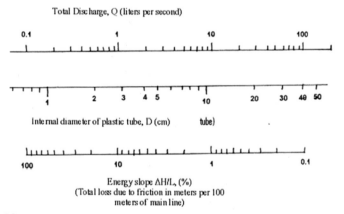

FIGURE 2.22   Nomograph for the design of main and secondary lines (MKS units).

2.4   DESIGN EXAMPLES

2.4.1   DESIGN OF LATERAL LINE ON UNIFORM SLOPE

2.4.1.1   DESIGN EXAMPLE NUMBER 1

The operational pressure of the lateral line is 6.5 psi (or 15 ft of head); the length of the lateral line is 300 ft; the total discharge is 2 gpm; the slope of the lateral line is of 2% (downward of the slope); and the size of the lateral line is 0.5 inches (ID = 0.625 inches). Review the acceptability of the design. One can use Figures 2.2 and 2.4 as follows:

Design procedure:

Step 1: Calculate $L/H = 300/15 = 20$.

Step 2: From Figure 2.2 (or Fig. 2.4) in the third quadrant, move vertically from $L/H = 20$ toward the line of total discharge ($Q = 2$ gpm) in the second quadrant; then establish a horizontal line toward the first quadrant.

Step 3: Move horizontally from $L/H = 20$ in the third quadrant to the line of slope of 2% in the fourth quadrant; then establish a vertical line toward the first quadrant.

Step 4: The point of intersection of two lines in the first quadrant shows a CU, $C_u = 97\%$ and variation in the dripper flow = 13%. The design is acceptable.

2.4.1.2 DESIGN EXAMPLE NUMBER 2

The length of the lateral line in a vegetable field is 150 ft and the slope of the line is 1% (downward of the slope). The dripper spacing is one feet. The dripper flow is 1 gph at an operational pressure of 15 psi. Design the size of the lateral lines.

Length of the lateral line, $L = 150$ ft
Operational pressure, $H = 15$ psi = $15/2.3 = 34$ ft of head.
Number of drippers = 150
Total discharge, $Q = 150$ gph for 150 drippers = 2.5 gpm.

Design procedure:

Step 1: Calculate $L/H = 150/34 = 4.4$

Step 2: In Figure 2.8 and for the desirable uniformity (zone A), determine the value of $\Delta H/L = 6$.

Step 3: In Figure 2.10, and for total discharge $Q = 2.5$ gpm and $\Delta H/L = 6$, determine the minimum size of the lateral line = 0.5 inches (ID).

2.4.1.3   DESIGN EXAMPLE NUMBER 2A

Assuming the same conditions indicated in the Example 2.4.1.2, design the size of the lateral line, for the following data:

Length of the lateral line   = 50 m
Operational pressure head   = 10 m
Number of drippers   = 600 1ph = 600/3600 = 0.167 lps.

Design procedure:

Step 1: Determine $L/H = 50/10 = 5$.

Step 2: In Figure 2.6, move vertically from L/H to the total discharge of the line $Q = 0.167$; then establish a horizontal line toward the first quadrant.

Step 3: Move horizontally from $L/H = 5$ to the line of the slope of 1% in the fourth quadrant; then establish a vertical line toward the first quadrant.

Step 4: The line of intersection is in the acceptable region. Therefore, the lateral line of 12 mm is used in the design.

2.4.1.4   DESIGN EXAMPLE NUMBER 3

The length of the lateral line in a vegetables field is 100 m; the slope of the lateral line is 1% toward underneath of the side. The dripper flow is 4 lph at an operational pressure head of 10 m. Design the size of the lateral line with the following data:

Length of the lateral line = 100 m
Operational pressure H = 10 m
Number of drippers   = 300
Total discharge Q = 4 × 300 = 1200 lph = 1200/3600 = 0.334 lps.

Design procedure:

Step 1: Determine $L/H = 100/10 = 10$.

Step 2: In Figure 2.9, move horizontally from $L/H = 10$ toward the line of 1% of slope in the fourth quadrant. From this point, establish a vertical line toward the first quadrant.

Step 3: Establish a point along this line in the first quadrant, in the upper margin of the desirable region. From this point, establish a horizontal line toward the second quadrant.

Step 4: Establish a vertical line in the second quadrant from $L/H = 10$, so as to intersect the horizontal line of the Step 3.

Step 5: For this point, determine $\Delta H/L = 3.5$ in the second quadrant.

Step 6: In Figure 2.11, for total discharge $Q = 0.334$ lps and $\Delta H/L = 3.5$: Determine the minimum size of the lateral line = 19 mm (or use size greater than 19 mm).

2.4.2 DESIGN OF THE LATERAL LINES ON NON-UNIFORM SLOPE

2.4.2.1 DESIGN EXAMPLE NUMBER 4

A lateral line of 0.5 inches (16 mm) diameter and 400 ft (122 m) of length is laid on a non-uniform slope. The non-uniform slope can be expressed as: 0–100 ft (0–30 m) = 3% of slope downward; 100–200 ft (30–60 m) = 2% of slope downward; 200–300 ft (60–90 m) = 0% of slope downward; 300–400 ft (90–120 m) = 3% of slope downward. The operational pressure head for the drip irrigation is 34 ft (or 10 m) and the total discharge for the lateral line is 2 gpm (0.13 lps). Assume that the total energy loss by friction is 5 ft (1.5 m). Review the variation in pressure with the following data:

H = 34 ft (10 m).
L = 400 ft (120 m).
ΔH = 5 ft (1.5 m) calculated by Eq 2.2 in Table 2.1.

Design procedure:

Step 1: Determine: $L/H = 11.76$ for British system or 12 for metric system; and
$\Delta H/H = 0.147$ for British system or 0.15 for metric system.

Step 2: Non-uniform slopes: plot $1/L$ vs. $\Delta H'_i/L$ in the first quadrant (Fig. 2.23).

Continue the procedure given in the previous section to revise the variation in pressure for four points and for a length ratio of 0.25, 0.50, 0.75, and 1.00, respectively (Table 2.2).

The results are given in Figure 2.23. It has been found in third quadrant that the variation in pressure along the lateral lines is less than 10%. The design is acceptable.

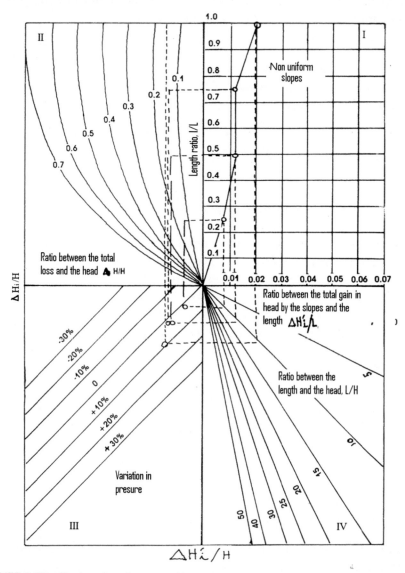

FIGURE 2.23   Design chart for non-uniform slope (with an example of design for Section 2.4.2).

TABLE 2.2   Length Ratio Versus Pressure Variation.

1/L	H_i (m)	$\Delta H'_i$ (m)	$\Delta H'_i/L$
0.25	3	0.9	0.0075
0.50	5	1.5	0.0125
0.75	5	1.5	0.0125
1.00	8	2.4	0.0200

2.4.3   DESIGN OF LATERAL LINE WITH DIFFERENT SIZES OF PIPES

2.4.3.1   DESIGN EXAMPLE NUMBER 5

In a papayas field, the length of lateral line is 1000 ft and is laid on a slope 5% downward. The papayas are planted in a zigzag pattern on both sides of the lateral line. The plant spacing is 5 ft. A micro tube of polyethylene is connected to the lateral line. Each spaghetti can deliver a flow of 2 gph at an operational pressure of 10 psi. Design the size of the lateral line, when different sizes of tubing of the lateral line can be used.

Design procedure:

Step 1: Total discharge, $Q = [(1000 \times 2)/5] \times [2]/60 = 13.33$ gpm.

Step 2: Slope of the lateral line = 5%

Step 3: If the lateral lines line is divided into 10 sections (Table 2.3), the size of the lateral line can be determined from Figure 2.14.

Step 4: Size of lateral line on a slope of 5% downward (Section 2.4.3.1):

TABLE 2.3   Size of Lateral Line.

Section	Discharge per section (gpm)	Size of lateral line (inches)
1	12.60	1.25
2	11.30	1.25
3	9.98	1.00
4	8.50	1.00
5	7.30	1.00
6	6.00	1.00

TABLE 2.3   *(Continued)*

Section	Discharge per section	Size of lateral line
	(gpm)	(inches)
7	4.66	0.75
8	3.33	0.75
9	2.00	0.50
10	0.67	0.50

2.4.3.2   DESIGN EXAMPLE NUMBER 6

Size of lateral line on a non-uniform slope (Section 2.4.3.2). Design the size of the lateral line. If the lateral line is laid on a non-uniform slope in Example 5 (downward the slope) as given below:

0–200 ft = 5%

200–400 ft = 3%

400–60 ft = 1%

600–80 ft = 3%

800–1000 ft = 5%

Design procedure:

Step 1: Total discharge, Q = 13.33 gpm.

Step 2: Slope of the lateral line = non-uniform.

Step 3: If the lateral line is divided into 10 sections, the size of the lateral lines can be determined from Figure 2.14. The results are given in Table 2.4.

TABLE 2.4   Size of Lateral Line for Example 6.

Section	Discharge per section	Slope	Size of lateral line
	(gpm)	(%)	(inches)
1	12.6	5	1.25
2	10.3	5	1.25
3	9.9	3	1.25
4	8.5	3	1.25
5	7.3	1	1.25
6	6.0	1	1.25

TABLE 2.3 *(Continued)*

Section	Discharge per section (gpm)	Slope (%)	Size of lateral line (inches)
7	4.7	3	1.00
8	3.3	3	0.75
9	2.0	3	0.75
10	0.7	5	0.50

2.4.4 DESIGN OF SECONDARY LINE USING AN INDIVIDUAL SIZE

2.4.4.1 DESIGN EXAMPLE NUMBER 7

The rectangular sugarcane field is one acre. The length of the secondary line is 100 ft and has a slope of zero. The secondary line supplies 15 gpm, at an operational pressure head of 34 ft. Design the size of secondary line.

Design procedure:

Step 1: Total discharge = 15 gpm

Step 2: Ratio of length of the secondary line and the operational pressure head, $L/H = 100/34 = 3$.

Step 3: Slope of the secondary line = 0.

Step 4: From Figure 2.8, $\Delta H/L$ is found to be 10% (variation in pressure of 20% for region A).

Step 5: In the Figure 2.10, the size of the line = 1 inch.

Step 6: Simplified design chart (Fig. 2.16) can also be used.

2.4.4.2 DESIGN EXAMPLE NUMBER 8

In Example 7, the secondary line is on a 5% slope (downward of the slope). Design the size of the secondary line.

Design procedure:

Step 1: Total discharge $Q = 15$ gpm.

Step 2: Ratio of length of the secondary line and the operational pressure head, $L/H = 100/34 = 3.3$.

Step 3: Slope of the secondary line: 5% downward.

Step 4: From the Figure 2.15, the size of the secondary line is approximated as one inch.

2.4.4.3   DESIGN EXAMPLE NUMBER 9

A secondary line of 10 m of length is installed in a vegetable field. The spacing between the lateral lines is one meter. These lateral lines are connected to the secondary line. The secondary line is laid on a zero slope. Design the size of secondary line with the following data:

Length of the secondary line, $L = 10$ m.

Operational pressure head, $H = 10$ m.

Total discharge for each lateral line, q = 0.167 lps.

Design procedure:

Step 1: Total discharge, $Q = 0.167 \times 10 = 1.67$ lps.

Step 2: Ratio of length of the secondary line and the pressure head, $L/H = 10/10 = 1.0$.

Step 3: Slope of the secondary line = 0.

Step 4: From Figure 2.18, the size of the secondary line is approximated as 25 mm.

2.4.5   DESIGN OF SECONDARY LINES OF DIFFERENT SIZES

2.4.5.1   DESIGN EXAMPLE NUMBER 10

In a sugarcane field, the secondary line is used for a rectangular subfield of two acres. The secondary line has a length of 300 ft and is laid on a slope 3% (downward the slope). Design the size of secondary line, when the engineer can design for different sizes.

Design procedure:

Step 1: Total discharge = 40 gpm.

Step 2: Slope = 3% uniform (downward).

Step 3: If the line secondary is divided into 10 sections (Table 2.5), the size of each section of the secondary line can be determined from Figure 2.14.

TABLE 2.5   Size of Secondary Line.

Section	Mean discharge (gpm)	Size of secondary line (inches)
1	38	2
2	34	2
3	30	2
4	26	1.5
5	22	1.5
6	18	1.5
7	14	1.5
8	10	1.25
9	6	1
10	2	3/4

2.4.6   DESIGN OF MAIN LINE

2.4.6.1   DESIGN EXAMPLE NUMBER 11

A drip irrigation system is designed for a papaya field of 50 acres (= 20 ha). The field is rectangular and is divided into subplots of one acre (0.4 ha). Each subplot is irrigated by a secondary line. The main line is located in the center of the field with 25 acres on each side. Each subplot is approximately 435 ft in length (130 m) and 100 ft wide (30 m). Each section of main line is 100 ft in length. The design capacity is 30 gpm (2 lps) for each subplot. There are 24 sections in total. At end of each section, there is an exit to supply 60 gpm (4 lps) for irrigating on both sides. The slopes of the main line are shown in Figure 2.19 or 2.20. The pressure of water required for the lateral line is 10 psi (pressure head of 17.5 m). Design the main line.

Design procedure:

Step 1: Plot slopes (profiles) of the main line as shown in Figure 2.19 or 2.20.

Step 2: Plot the required pressure (10 psi or 17.5 m) along the main line as shown in Figure 2.19 or 2.20 = 23 ft above the ground.

Step 3: Determine the energy slope. From Figures 2.19 and 2.20, the energy slope is determined as 1%. The size of main line can be determined from Figure 2.21 or 2.22, utilizing 1% of the energy gradient line. The results are shown in Table 2.6.

TABLE 2.6   Sizes of the Main Line as Determined from Figure 2.21 or 2.22.

Section	British units		Metric units	
	Discharge (gpm)	Diameter (inches)	Discharge (lps)	Diameter (cm)
0	1500	–	48	25
1	1440	10	46	25
2	1380	10	44	25
3	1320	10	42	20
4	1260	8	40	20
5	1200	8	38	20
6	1140	8	36	20
7	1080	8	34	20
8	1020	8	32	20
9	960	8	30	20
10	900	8	28	20
11	840	8	26	20
12	780	8	24	20
13	720	8	22	20
14	660	8	20	15
15	600	6	18	15
16	540	6	16	15
17	480	6	14	15
18	420	6	12	15
19	360	6	10	15

TABLE 2.6 *(Continued)*

Section	British units		Metric units	
	Discharge (gpm)	Diameter (inches)	Discharge (lps)	Diameter (cm)
20	300	5	8	12.5
21	240	5	6	12.5
22	180	4	4	10
23	120	4	2	10
24	60	3		

There is an outlet at the entrance of the section one to irrigate the plots on both sides of section one.

KEYWORDS

- blasius formula
- coefficient of uniformity
- control system
- drip irrigation system
- dripper
- energy gradient line
- GPM
- British units of volumetric flow
- gpm
- Hhdraulic gradient
- lateral line
- LPS
- metric units of volumetric flow
- main line
- orifice
- pipe or tube
- polyethylene
- secondary line
- LPH
- smooth surface
- spaghetti

REFERENCES

1. Howell, T. A.; Hiler, E. A. Designing Trickle Irrigation Laterals for Uniformity. *J. Irrig. Drain. Div., ASCE.* Proc. Paper 10983. **1974,** *100*(IR4), 443–454.
2. Howell, T. A.; Hiler, E. A. Trickle Irrigation Lateral Design. *Trans. Am. Soc. Agric. Eng.* **1974,** *17*(5), 902–908.
3. Williams, G. S.; Hazen, A. *Hydraulic Tables.* 3rd ed. John Willey and Sons: New York, 1960.
4. Wu, I. P. Design of Drip Irrigation Main Lines. *J. Irrig. Drain. Div., ASCE.* Proc. Paper 11803. **1975,** *101*(IR4), 265–278.
5. Wu, I. P.; Gitlin, H. M. Hydraulics and Uniformity for Drip Irrigation. *J. Irrig. Drain. Div., ASCE.* Proc. Paper 9786. **1973,** *99*(IR3), 157–168.
6. Wu, I. P.; Gitlin, H. M. Hydraulics Irrigation Design Based on Uniformity. *Trans. Am. Soc. Agric. Eng.* **1973,** *17*(3), 157–168.
7. Wu, I. P.; Gitlin, H. M. *Design of Irrigation Lines*; Technical Bulletin No.96: Hawaii Agricultural Experimental Station, University of Hawaii, USA, 1974.
8. Wu, I. P.; Gitlin, H. M. Energy Gradient Line for Drip. *Irrig. Drain. Div., ASCE.* Proc. Paper 11750. **1975,** *101*(IR4), 323–326.
9. Wu, I. P.; Gitlin, H. M. *Drip Irrigation Designs on Non-Uniform Slopes.* Paper presented at the 1975 Winter Meeting of American Society of Agricultural Engineers, Chicago, IL, 1975.
10. Wu, I. P.; Gitlin, H. M. Design Drip Irrigation Lines Varying Pipes Sizes. *J. Irrig. Drain. Div.,* ASCE. Proc. Paper 13384. **1977,** *103*(IR4), 499–503.
11. Wu, I. P.; Gitlin, H. M. *Drip Irrigation Systems Design*; Bulletin No. 144 and 156 of the Cooperative Extension Service, University of Hawaii: USA, 1975.

CHAPTER 3

SELECTED DESIGN EXAMPLES OF DRIP IRRIGATION SYSTEMS

VISHAL K. CHAVAN

Assistant Professor and Senior Research Fellow in SWE, AICRP for Dryland Agriculture, Dr. PDKV, Akola, Maharashtra, India.

E-mail: vchavan2@gmail.com

CONTENTS

ABSTRACT

In this chapter, design examples for drip irrigation system have been presented for orange, guava, banana, pomegranate, cauliflower, and sugarcane crops under Indian conditions.

3.1   INTRODUCTION

Indian Agriculture mainly depends upon monsoon rain, which is unevenly distributed and not adequate to meet the moisture requirement of the crops for successful farming. India with 2.4% of the world's total area has 16% of the world's population, but has only 4% of the total available fresh water. India has a capacity to store about 200 billion cubic meters of water, a gross irrigated area of about 90 million hectares and an installed hydropower capacity of about 30,000 megawatts. The estimates by Government of India and ministry of water resources indicates that in 2050 India will need to increase by five times more water supplies to industries, and 16 times more for energy production, while its drinking water demand will double, and irrigation demand will rise by 50 percent. A United Nations task force on water predicted that by 2025, 3 billion people will face "water stress" conditions, lacking enough water to meet all human and environmental needs. This clearly indicates the need for water resource development, conservation, and optimum use. The water resources of the most of the regions in India are limited and water is applied by various methods. Thus, irrigation is necessary in one form or another.

Drip irrigation, also called trickle irrigation or micro irrigation, is a localized irrigation method that slowly and frequently provides water directly to the plant root zone. It is known as a low cost water delivery system. However, not until the innovation of polyethylene plastics in the 1960s did drip irrigating begin to gain momentum. Traditionally, irrigation had relied upon a broad coverage of water to an area that may or may not contain plants. Promoted for water conservation, drip irrigation does just the opposite. It applies small amounts of water (usually every two or three days) to the immediate root zone of plants. In drip irrigation, water is delivered to individual plants at a low pressure and delivery rate to specific areas or zones in the landscape or garden. The slow application promotes a thorough penetration of the water to individual plant root zones and reduces potential runoff. The depth of water penetration depends on the length of time the system is allowed to operate and the texture of the soil.

Drip irrigation method was found to increase the crop yield by 20–90% and reduce water use by 30–70% for different crops. The suitability of any irrigation system mainly depends upon its design, layout, and performance. Due to its merits and positive effects, drip irrigation has been rapidly popular in India and also the state governments are promoting drip irrigation on a large scale by providing subsidy. The advantage of using a drip irrigation system is that it can significantly reduce soil evaporation and increase water use efficiency by creating a low, wet area in the root zone. Due to water shortages in many parts of the world today, drip irrigation is becoming quite popular [1, 2, 3]. In 2000, more than 73% of all agricultural fields in Israel were irrigated using drip irrigation systems, and 3.8 million hectares worldwide were irrigated using drip irrigation systems.

Due to limited water resources and environmental consequences of common irrigation systems, drip irrigation technology is getting more attention and playing an important role in agricultural production, particularly with high value cash crops such as greenhouse plants, ornamentals, and fruit. Therefore, use of drip irrigation systems is rapidly increasing around the world. Despite its advantages, in drip irrigation system, emitter clogging is one of the major problems which can cause large economic losses to the farmers. Emitter clogging is directly related to the quality of the irrigation water, which includes factors such as suspended solid particles, chemical composition and microbes, and also insects and root activities within and around the tubing can also cause problems. The major operational difficulties in drip irrigation method arise from the clogging of dripper which reduces the efficiency and crop yield.

Emitter clogging continues to be a major problem in micro irrigation systems. For high-valued annual crops and for perennial crops, where the longevity of the system is especially important, and emitter clogging can cause large economic losses. Even though information is available on the factors causing clogging, control measures are not always successful. Problems can be minimized by appropriate design, installation, and operational practices. Reclamation procedures to correct clogging increase maintenance costs, and unfortunately, may not be permanent. Clogging problems often discourage the operators, and consequently cause the abandonment of the system and the return to a less efficient irrigation application method.

Emitter clogging is directly related to the quality of the irrigation water, which includes factors such as suspended particle load, chemical composition, and microbial type and population. Insect and root activities within and around the tubing can cause similar problems. Consequently, these factors dictate the type of water treatment or cultural practices necessary for clogging prevention. Clogging problems are often site-specific and solutions are not always available or economically feasible. No single foolproof quantitative method is available for estimating clogging potential. However, by analyzing the water for some specific constituents, possible problems can be anticipated and control measures can be formulated.

Most tests can be made in the laboratory. However, some analyses must be made at the sampling sites because rapid chemical and biological changes can occur after the source water is introduced into the micro irrigation system. Water quality can also change throughout the year so that samples should be taken at various times over the irrigation period. These are further rated in terms of an arbitrary clogging hazard ranging from minor to severe. Clogging problems are diminished with lower concentrations of solids, salts, and bacteria in the water. Additionally, clogging is aggravated by water temperature changes.

The causes of clogging differ based on emitter dimension and positions in lateral found. The tube emitter system with laminar flow suffers more severe clogging than the labyrinth system with turbulent flow, because laminar flow is predisposed to clogging.

Emitter clogging has often been recognized as inconvenient and one of the most important concerns for drip irrigation systems, resulting in lowered system performance and water stress to the non-irrigated plant. Partial and total plugging of emitters is closely related to the quality of the irrigation water, and occurs as a result of multiple factors, including physical, biological, and chemical agents. Favorable environmental conditions in drip irrigation systems can cause rapid growth of several species of algae and bacteria resulting in slime and filament buildup, which often become large enough to cause biological clogging. On the other hand, some of the bacterial species may cause emitter clogging due to the precipitation of iron, manganese, and sulfur minerals dissolved in irrigation water. Filtration, chemical treatment of water and flushing of laterals are means generally applied to control emitter clogging. Physical clogging can be eliminated with the use of fine filters and screens. Emitter clogging is directly related to irrigation water quality, which appears a function of the amount of suspended solids, chemical constituents of water and microorganism activities in water. Therefore, the mentioned factors have a strong influence on the precautions that will be taken for preventing the plugging of the emitters. During irrigation, some clogging due to micro-organism activities take place in cases when wastewater is used.

In micro-irrigation systems that are characterized by a number of emitters with narrow nozzles, irrigation uniformity can be spoilt by the clogging of the nozzles with particles of chemical character. Chemical problems are due to dissolved solids interacting with each other to form precipitates, such as the precipitation of calcium carbonate in waters rich in calcium and bicarbonates. In locations where the amount of the ingredients as dissolved calcium, bicarbonate, iron, manganese, and magnesium are excessive in irrigation water, the emitters are clogged by the precipitation of these solutes. Chemical precipitation can be controlled with acid injection. However, biological clogging is quite difficult to control. Chlorination is the most common practice used in the prevention and treatment of emitter clogging caused by algae and bacteria. Calcium hypochlorite, sodium hypochlorite, and particularly chlorine are the most common and inexpensive treatments for bacterial slimes and for inhibition of bacterial growth in drip irrigation systems. However, continuous chlorination would increase total dissolved solids in the irrigation water and would contribute to increased soil salinity. Drip irrigation is being widely accepted in the world and becoming increasingly popular in areas with water scarcity and salt problems. In regions where good quality water is either insufficient or not available, saline water irrigation is inevitable; thus drip irrigation has gained immense value in some specific situations with not only good quality water but saline water, too. Drip irrigation is being used in many areas with surface water too, that is, water from streams/canals and rivers which contain various sizes of sand particles and sand concentrations. Drip irrigation is based on a very slow and frequent application of water from relatively small nozzles or orifices, which discharge 1–10 lph. In order to achieve such a small discharge, the passage and orifices must be very small. In most cases, the diameter of an emitter orifice is less than 1 mm understandably; a formidable obstacle to the successful operation of the system over its intended life of service is clogging of emitters. Greater water application uniformity is one of the significant advantages that a properly designed and maintained drip system can offer over other methods of irrigation. In many cases, the yield of crops may be directly related to the uniformity of water application. Partial or complete clogging drastically affects water application uniformity and, hence, may put a complete system out of operation, causing heavy loss to the crop and damage to the system itself. Thus, emitter clogging can nullify all the advantages of drip irrigation.

3.2 DESIGN OF DRIP IRRIGATION SYSTEM FOR SWEET ORANGE

1. Size of land — : 200 m × 500 m
2. Type of soil — : Black type with well drain property
3. Land slope — : Uniform slope
4. Maximum evaporation — : Bore-Well situated at center of field
5. Water source — : 10.5 mm/day
6. Available discharge — : 6 Ips
7. Static head — : 10 m
8. Wetted area for orange — : 20%
9. Life of orchard — : One year old
10. Variety — : Marmalade orange
11. Spacing — : 6 × 6 m

3.2.1 DESIGN PROCEDURE

3.2.1.1 NET DEPTH OF WATER

Evapotranspiration of crop (ET) = PE × Kp × Kc = 10.5 × 0.8 × 0.75 = 6.30 mm/day

Total volume of water required = Plant spacing × Row spacing (m) × Wetted area (in fraction) × Depth of water (cm) = 6 × 6 × 0.20 × 6.30 = 45.36 l/day/plant.

Consider 90% emission uniformity of the drip system.

Volume of water required = 45.36/0.90 = 50.40 l/day/plant (% wetted area is the area, which is shaded due to its canopy cover, when the sun of overhead, which depends on the stage of crop growth).

3.2.1.2 EMITTER SELECTION AND POSITIONING

The planting of orange is for one-year-old orchard and the soil is black and well drained: Use two drippers each of 8 lph discharge rate.

Operation time of system (T_0)

$$= \frac{volume\ of\ water\ to\ be\ applied\ /\ tree}{total\ dripper\ discharge} = 50.40/16 = 3.15 \text{ hours} = 3 \text{ h } 9 \text{ min.}$$

3.2.2 LAYOUT OF SYSTEM

1. Length of lateral: 50 m
2. Spacing between two successive emitters = From the center of the trunk 50 cm on either side: 1 m and between two trees; distance between two emitters is 5 m.
3. Number of trees per lateral: 50/6 = 8.33, say 8
4. Number of drippers per lateral: 8 × 2 =16 at the rate of two drippers per tree
5. Total number of laterals = According to layout the total number of 50 m lateral = 250/6 = 41.66, say 42 (250 m is the length of each manifold. On one manifold on either side = 42 + 42 = 84 No. of laterals.
   Therefore the total No. of laterals on four manifolds = 4 × 84 = 336 No. of laterals each 50 m in length.
6. Total number of emitters: There are 16 emitters on one lateral. Therefore, 336 × 16 = 5376 No. of emitters.
7. Total length of lateral = 336 × 50 = 16,800 m
8. Discharge through one lateral = 16 × 8 = 128 lph

3.2.2.1 NUMBER OF SHIFTS

Discharge through one lateral = 16 drippers × 8 lph = 128 lph

$$\text{No of laterals can be operated at a Hme} = \frac{Discharge\ available\ per\ h}{Discharge\ through\ one\ lateral}$$

$$= \frac{6 \times 3600}{128} = 168.75 = 168 \text{ number of laterals can be operated at a time.}$$

Therefore, we have to split the field into four units and operate accordingly.

3.2.2.2 PERMISSIBLE HEAD LOSS DUE TO FRICTION (J)

Taking into account 10% variation = 10 m × 0.10 = 1 m (for 8 lph dripper operated at 10 m head). Hazen–William's equation is used for determining the head loss due to friction = $J = 1.526 \times 10^4 (Q/C)^{1.852} D^{-4.87} (L + Le) F$

Where: J = Head loss due to friction in m, Q = Flow rate in m³/h, C = Constant depending on pipe material, D = Inside pipe diameter in cm, L = Length of pipe or tubing in m, Le = Increase in length of pipe or tubing due to connections of emitters in m; F = Reduction coefficient. Use f = 0.38 for 16 outlets on lateral, L = 50 m, Le = No. of emitters × factor = 16 × 0.39 = 6.24 m, C = 30, and assume D = 12 mm. Therefore: J = 1.526 × 10^4 $(0.128/130)^{1.852}$ × $(1.2)^{-4.87}$ (50 + 6.24) × 0.38 = 0.37 m, which is less than 1 m, therefore the lateral of 12 mm size is sufficient to meet the hydraulic requirements.

3.2.2.3   SIZE OF SUBMAIN

Head loss at the outlet of each submain = 10 + 0.37 = 10.37 m

Number of submains = 4, Length of submain = 250 m

Discharge (Q) through one submain = Q_{lat} × No. of laterals = 128 × 84 = 10, 752 lph or Qs = 10.752 m³/h.

For 84 outlets, F = 0.36, L – 250 m, Le = 84 × 0.36 = 30.24 m, C = 140

Now assume D = 63 mm,

J = 1.526 × 10^4 $(10.752/140)^{1.852}$ × $(6.3)^{-4.87}$ (250 + 30.24) × 0.36 = 1.69 m

Head at the inlet of submain = Hsubmain + Hlat + Hemitter + H slope

$\qquad$ = 1.69 + 0.37 + 10.00 + 0.00 = 12.06 m

Head loss in the submain = 12.06 − 10.37 = 1.69 m

Variation in head = $\dfrac{1.69}{12.06}$ × 100 = 14.01%

This variation in head loss is less than 15%. Hence, the size of 50 mm PVC pipe is acceptable for sub mains.

3.2.2.4   SIZE OF MAIN

Length of main = 50 m

Total discharge of main = discharge of submain = 10.752 m³/h, Assume D = 63 mm,

J = 1.526 × 10^4 $(10.752/140)^{1.852}$ × $(6.3)^{-487}$ (50 + 1) × 1 = 0.85 m

Head at the inlet of main = 12.06 + 0.85 = 12.91 m

Variation in heads = $\dfrac{0.85}{12.91}$ × 100 = 6.58%

As the variation in head does not exceed 15%, 63 mm diameter of main is acceptable.

3.2.2.5 TOTAL DYNAMIC HEAD (TDH)

1. Pressure head at emitter: 10.00 m
2. Head loss in lateral: 0.37 m
3. Head loss in submain: 1.69 m
4. Head loss in main line: 0.85 m
5. Static head: 10.00 m
6. Head losses due to filters: 7.00 m
7. Fertilizer injection: 1.00 m
8. Main control valve: 0.50 m
9. Head losses due to other : 3.14 m
10. Fittings at the rate of 10% for safety factor
    TDH = 34.55 m

3.2.2.6 PUMP SIZE (HP)

$$H.P = \frac{QH}{75n}$$

Where: Q = Discharge of mainline in Ips, H = Total dynamic head in m, n = Efficiency of pump = 65%.

$$H.P = \frac{\left(\frac{10752}{3600}\right) \times 34.55}{75 \times 0.65} = 2.11 \text{ or use 2. Use 2 hp pump for running a drip}$$

system for 10 ha of orange.

3.2.2.7 DETAILS OF DRIP UNIT REQUIRED

1. Drippers: #5376, 8 lph, two drippers per tree.
2. Laterals: 336 No., each 50 m, total length = 16,800 m, dia. 12 mm, LLDPE
3. Submain: 4 No., total length = 1000 m, dia. 63 mm, PVC
4. Main: 100 m in length, dia. 63 mm, PVC
5. Pump: 2 hp

6. Sand and screen filter coupled with by-pass and back flush system.
7. Fittings and accessories: tees, bends, control, values, grommet and take-off, and flushings, caps, cement solution, pressure gages etc.

The cost of the drip irrigation system can be estimated as per the components estimated by considering the current market rates.

3.3   DESIGN OF DRIP IRRIGATION FOR GUAVA

1. Size of land: 200–400 m
2. Type of soil: medium to light
3. Land slope: Uniform slope
4. Maximum evaporation: 11 mm/d
5. Water source: Well situated at west-south corner of field
6. Available discharge: 5.5 Ips
7. Static head: 10 m
8. Wetted area for Guava: 20%
9. Life of orchard: One year old
10. Crop type & variety: Guava var. Sardar
11. Total design area: 8 ha
12. Tree spacing: 4 × 4 m

3.3.1   DESIGN PROCEDURE

3.3.1.1   NET DEPTH OF WATER

Evapotranspiration of crop (ET) = PE × Kp × Kc = 11 × 0.7 × 0.75 = 5.77 mm/day

3.3.1.2   VOLUME OF WATER

Total volume of water required = Plant spacing × Row spacing (m) x Wetted area (in fraction) – Depth of water (cm) = 4 × 4 × 0.20 × 5.77 = 18.46 l/day/plant

Consider 90% emission uniformity of the drip system. Volume of water required = 8.46/0.90 = 20.51 l/day/plant.

3.3.1.3 EMITTER SELECTION AND POSITIONING

As the planting of guava is of one year old and the soil is medium to light. Therefore, 1dripper, each of 4 lph discharge, is sufficient to deliver the volume of 20.51 l/day/plant.

3.3.1.4 OPERATION TIME OF SYSTEM (T_o)

Operation time (To) = $\dfrac{volume\ of\ water\ to\ be\ applied\ /\ tree}{total\ dripper\ discharge} = \dfrac{20.51}{4} = 5.12\,h$
= 5 h 7 min.

3.3.1.5 LAYOUT OF SYSTEM

1. Length of lateral = 50 m
2. Spacing between two successive emitters = 4 m
3. Number of trees per lateral: 50/4 = 12.5 or use 13
4. Number of drippers per lateral = 13
5. The number of laterals on one manifold for 50 m each lateral = 400/4 = 100
6. Use one manifold on either side, No. of laterals = 100 + 100 = 200 laterals. Therefore, the total number of laterals on 2 manifold = 200 × 2 = 400 No. of 50 m in length.
7. Total number of emitters: There are 13 emitters on one lateral = 400 × 13 = 5200 emitters.
8. Total length of lateral = 400 × 50 = 20,000 m
9. Discharge through one lateral = 13 × 4 = 52 lph

3.3.1.6 NUMBER OF SHIFTS

Discharge through one lateral = 13 drippers × 4 lph = 52 lph

No of laterals can be operated at a time = $\dfrac{Discharge\ available\ per\ hr}{Discharge\ through\ one\ lateral}$ = $\dfrac{5.5 \times 3600}{52}$ = 380.76 = 380 No. of laterals can be operated at a time.

However, we have to split the whole field into two units and operate accordingly as per the layout considering that there are 400 laterals measuring 50 m length.

3.3.1.7   SIZE OF LATERAL

The permissible head loss, due to friction by taking into account 10% head variation, will be 10 m × 0.10 = 1 m (for 4 lph dripper operated at 10 m head). The Hazen–William's equation is used for determining the head loss due to friction:

$$J = 1.526 \times 10^4 (Q/C)^{1.852} D^{-4.87} (L + Le) F$$

In this example: f = 0.4 for 13 outlets on lateral, L = 50 m, Le = No of emitters × factor = 13 × 0.4 = 5.2 m, C = 30, and assuming D = 12 mm, we get:

$$J = 1.526 \times 10^4 (0.052/130)^{1.852} \times (1.2)^{-487} (50 + 5.2) \times 0.4 = 0.07 \text{ m},$$

which is than 1 m. Therefore, lateral of 12 mm in size is sufficient to meet the hydraulic requirements.

3.3.1.8   SIZE OF SUBMAIN

Head loss at the outlet of submain = 10 + 0.07 = 10.07 m

Number of submains = 2, Length of each submain = 400 m

Discharge (Q) through one submain = Q_{lat} × No. of laterals = 52 × 200 = 10,400 lph or Qs = 10.400 m³/h.

For 200 outlets F = 0.36, L = 400 m, Le = 200 × 0.36 = 72 m, C = 140, and assuming D = 75 mm we get: $J = 1.526 \times 10^4 (10.400/140)^{1.852} \times (7.5^{-4.87} (400 + 72) \times 0.36 = 1.15$ m

Head at the inlet of submain = Hsubmain + Hlat + Hemitter + Hslope = 1.15 + 0.07 + 10.00 + 0.00 = 11.22 m

Head loss in the submain = 11.22 − 10.07 = 1.22 m

Variation in head = $\dfrac{1.22}{11.22}$ × 100 = 10.88%. This variation in head loss is less than 20%. Hence, the size of 75 mm PVC pipe can be accepted for submain.

3.3.1.9   SIZE OF MAIN

Length of main = 150 m

Total discharge for main = discharge for all submains = 10.400 m³/h

Now, assume D = 75 mm,

$$J = 1.526 \times 10^4 (10.400/140)^{1.852} \times (1.5)^{-4.87} (150 + 1) \times 1 = 1.02 \text{ m}$$

Head at the inlet of main = 11.22 + 1.02 = 12.24 m

Variation in heads = $\dfrac{1.02}{12.24} \times 100 = 8.33\%$ = This variation in head is not exceeding 20%, Therefore, 75 mm dia of main can be accepted.

3.3.1.10   TOTAL DYNAMIC HEAD (TDH)

1. Pressure head at emitter = 10.00 m
2. Head loss in lateral = 0.07 m
3. Head loss in submain = 1.15 m
4. Head loss in main line = 1.02 m
5. Static head = 10.00 m
6. Head loss due to filters = 7.00 m
7. Head loss due to fertilizer injection = 1.00 m
8. Head loss due to control valves = 0.50 m
9. Head loss due to fittings etc. = 3.07 m and use 10% for safety factor
    Therefore, TDH = sum of all = 33.81 m

3.3.1.11   PUMP SIZE (HP)

$$Hp = \frac{QH}{75n} = \frac{\left(\dfrac{10400}{3600}\right) \times 33.81}{75 \, x \, 0.65} = 2.01 \text{ or use 2 hp.}$$

Use a pump of 2 hp for running a drip system for 8 ha of guava.

3.3.1.12   DETAILS OF DRIP UNIT

1. Drippers: 5200 at the rate of 4 lph each dripper and one dripper per guava tree.
2. Laterals : 400, each 50 m in length, total length = 20,000 m, dia. 12 mm, LLDPE
3. Submains: 2, total length = 800 m, dia 75 mm, PVC
4. Main: 150 m in length, dia. 75 mm, PVC
5. Pump: 2.0 hp
6. Sand and screen filter coupled with by-pass and back flush system. Fittings and accessories: tees, bends, control valves, grommet, and take-off, flushing caps, cement solution, pressure gages etc.

7. The cost of the drip irrigation system can be estimated as per the components estimated by considering the current market rates.

3.4 DESIGN OF DRIP IRRIGATION SYSTEM FOR GRAPES

1. Size of land: 100 m × 100 m
2. Type of soil: medium type
3. Land slope: 0.30% (S-N)
4. Maximum evaporation: 12.5 mm/day
5. Water source: well situated east-south corner
6. Available discharge: 3.5 lps
7. Static head: 10 m
8. Wetted area for grape: 50%
9. Spacing: 3 m × 3 m
10. Crop type: Grape (var. Sharad seedless) on 1 ha of land.

3.4.1 DESIGN PROCEDURE

3.4.1.1 NET DEPTH OF WATER

Evapotranspiration of crop (ET) = PE × Kp × Kc = 12.5 × 0.7 × 0.75 = 6.5 6 mm/day

3.4.1.2 VOLUME OF WATER

Total volume of water required = Plant spacing × row spacing (m) × wetted area (in fraction) × Depth of water (cm) = 3 × 3 × 0.30 × 6.56 = 17.71 l/day/plant.

Considering 90% emission uniformity of the drip system, we get:
Volume of water required = 17.71/0.90 = 19.68 l/day/plant.

3.4.1.3 EMITTER SELECTION AND POSITIONING

The age of grape tress is 1 year (the trees are grown fully). The soil is medium. Use one dripper per tree, each dripper of 4 lph.

3.4.1.4 OPERATION TIME OF SYSTEM (T_o)

Operation time (To) = $\dfrac{\text{volume of water to be applied / tree}}{\text{total dripper discharge}} = \dfrac{19.68}{4} = 4.92\,h$
or 4 h 55 min.

3.4.1.4 LAYOUT OF SYSTEM

1. Length of lateral = 50 m
2. Number of trees per lateral = 16.66 or 17
3. Number of drippers per lateral = 17 drippers
4. Number of laterals = on one side of manifold of 100 m =33.33 or 34
5. Total number of laterals on either side of manifold of 100 m length = 34 + 34 = 68
6. Total number of dripper/ha = 68 × 17 = 1156
7. Total length of lateral = 50 m × 68 laterals = 3400 m

3.4.1.5 NUMBER OF SHIFTS

Discharge through one lateral = 17 drippers × 4 lph = 68 lph
    Number of laterals that can be operated at a time

$= \dfrac{\text{Discharge available per hr}}{\text{Discharge through one lateral}} = \dfrac{3.5 \times 3600}{68} = 185.29 \text{ or } 185 \text{ of laterals}$

can be operated at a time. Therefore, all 68 laterals can be operated at one time. The excess water in the system can be diverted into the well using by-pass system or can be used for other farm activities.

3.4.1.6 SIZE OF LATERAL

The permissible head loss due to friction, by taking into account 10% variation in head, will be 10 m × 0.10 = 1 m (For 4 lph dripper operated at 10 m head). The Hazen–Williams equation used for determining the head loss due to friction is:

$J = 1.526 \times 10^4 (Q/C)^{1.852} D^{-4.87} (L + Le) F$

Use F = 0.39 for 17 outlet on laterals, L = 50 m, Le = No of emitters x factor = 17 × 0.39 = 6.63 m, C = 130 and assuming D = 12 mm, we get:

$J = 1.526 \times 10^4 (0.068/130)^{1.852} (1.2)^{-4.87}(50 + 6.63) 0.36 = 0.1072$ m, which is less than 1 m. Therefore, lateral of 12 mm size is sufficient to meet the hydraulic requirements.

3.4.1.7   SIZE OF SUBMAIN

Head loss at the outlet of submain = 10 + 0.1072 = 10.10 m

Number of submains = 1

Length of submain = 100 m

Discharge(Q)throughsubmain=Qlateral×No.oflaterals=68×68=4624Iphor 4.624/m³/h. For 68 outlets F = 0.36, L = 50 m, Le = 68 × 0.36 = 24.48 m, C = 140, and assuming D = 50 mm or 5.0 cm, we get:

$J = 1.526 \times 10^4 (4.624/140)^{1.852} (5.0)^{-4.87}(100 + 24.48) 0.36 = 0.4875$ m

Head at the inlet of submain = Hsubmain + Hlateral + Hemitter + Hslope = 0.4875 + 0.10 + 10 + 0.22 = 10.80 m, where Hslope = 0.75 × 0.30 = 0.22

Head loss in the submain = 10.80− 10.10 = 0.7075 m

Variation in head = 0.7075/10.80 = 6.55%, which is less than 20%. Hence the size of 50 mm PVC pipe can be accepted for sub main.

3.4.1.8   SIZE OF MAIN

Length of main = 50 m

Total discharge of main = Discharge of submain = 4.624 m³/h

C = 140, f = 1 for one outlet, Le = 1 × 4 = 1, and assuming D = 50 mm, we get:

$J = 1.526 \times 10^4 (4.624/140)^{1.852} \times 5^{-4.87}(100 + 1) \times 1 = 1.0987$ m

Head at the inlet of main = 10.80 + 1.09 = 11.89 m

Variation in head = $\dfrac{(11.89 - 10.80)}{11.89} \times 100 = 9.17\%$ which is less than 20%.
Therefore 50 mm of main can be used.

3.4.1.9   TOTAL DYNAMIC HEAD (TDH)

1.   Pressure head at emitter = 10.00 m

2. Head loss in lateral = 0.1072 m
3. Head loss in submain = 1.487 m
4. Head loss in main line = 1.09 m
5. Static head = 10.00 m
6. Head loss due to filters = 7.00 m
7. Head loss due to fertilizer injection = 1.00m
8. Head loss due to main control valve = 0.50 m
9. Head loss due to fittings etc. = 3.01 m
   TDH = sum of all = 33.19 m

3.4.1.10 PUMP SIZE (HP)

$$\text{H.P} = \frac{QH}{75n} = \frac{\left(\dfrac{4642}{3600}\right) \times 33.19}{75 \times 0.65} = 0.87 \text{ hp. Use 1 hp pump for running a drip}$$

system for grape.

3.4.1.11 DETAILS OF DRIP UNIT REQUIRED

1. Drippers: 1156 with 4 lph
2. Laterals: 68 each 50 m, total length = 3400 m, dia 12mm, LLDPE
3. Submain: one of 100 m, dia 50 mm, PVC
4. Main: 50 m in length, dia 50mm, PVC
5. Pump: 1 hp
6. Sand and screen filter coupled with by-pass and back flush system.
7. Fittings and accessories: tees, bends, control, values, grommet and take-off, and flushing caps, cement solution, pressure gages etc.

3.5   DESIGN OF DRIP IRRIGATION SYSTEM FOR POMEGRANATE

1. Size of land: 100 m × 150 m
2. Type of soil: light textured
3. Land slope: uniform slope
4. Maximum evaporation: 12.0 mm/day
5. Water source: Well situated east– north corner
6. Available discharge: 4 lps
7. Static head: 10m

8. Wetted area for grape: 20%
9. Life of orchard: newly planted
10. Crop type: pomegranate (Bhagawa) on 1.5 ha of an orchard
10. Tree spacing: 5 m × 4 m

3.5.1 DESIGN PROCEDURE

3.5.1.1 NET DEPTH OF WATER

Evapotranspiration of crop (ET) = PE × Kp × Kc = 12.0 × 0.7 × 0.75 = 6.3 0 mm/day

3.5.1.2 VOLUME OF WATER

Total volume of water required = Plant spacing × Row spacing (m) × Wetted area (in fraction) × Depth of water (cm) = 4 × 5 × 0.20 × 6.30 = 25.20 l/day/plant.

Considering 90% emission uniformity of the drip system, volume of water required = 28.00 l/day/plant.

3.5.1.3 EMITTER SELECTION AND POSITIONING

The planting of pomegranate is new and the soil is light textured. Use two drippers, each of 4 lph discharge. The total discharge rate = 8 lph for two emitters (4 × 2).

3.5.1.4 OPERATION TIME OF SYSTEM

Operation time (To) = $\dfrac{volume\,of\,water\,to\,be\,applied\,/\,tree}{total\,dripper\,discharge}$ = $\dfrac{28}{8}$ = 3.50 h = 2 h 30 min.

3.5.1.5 LAYOUT OF SYSTEM

1. Length of lateral = 50 m
2. Number of trees per lateral: = 50/4 = 2.5 or 13

3. Number of drippers per lateral = 13 × 2 = 26 at the rate of 2 drippers per tree
4. Total number of laterals (150 m length/5 m row to row) = 30 + 30 = 60 laterals on either side of manifold.
5. Total number of emitters = 26 × 60 = 1560
6. Total length of lateral = 50 m × 60 laterals = 3000 m

3.5.1.6 NUMBER OF SHIFTS

Discharge through one lateral = 26 drippers × 8 lph = 104 lph
  Number of laterals that can be operated at a time =

$$\frac{Discharge\ available\ per\ hr}{Discharge\ through\ one\ lateral}\quad \frac{4\times3600}{104} = 138.46\ or\ 139.\ All\ laterals$$

can be operated at a time. By-pass system is necessary.

3.5.1.7 SIZE OF LATERAL

The permissible head loss due to friction, by taking into account 10% head variation, will be 10 m × 0.10 = 1 m (For 4 lph dripper operated at 10 m head). The Hazen–William equation used for determining the head loss due to friction is:

$$J = 1.526 \times 10^4\,(Q/C)^{1.852}\,D^{-4.87}\,(L + Le)\,F$$

Using f = 0.38 for 26 outlets on lateral, L = 50 m, Le = No of Emitters x factor = 26 × 0.38 = 9.88 m, C = 130, and assuming D = 12 mm we get:

$$J = 1.526 \times 10^4\,(0.104/130)^{1.852} \times (1.2)^{-4.87}\,(50 + 9.88) \times 0.38 = 0.26\ m,$$

which is less than 1 m. Therefore lateral of 12 mm size is sufficient to meet the hydraulic requirements.

3.5.1.8 SIZE OF SUBMAIN

Head loss at the outlet of submain = 10 + 0.26 = 10.26 m

Number of submain = 1

Length of submain = 150 m

Discharge (Q) through submain = Qlat × No. of laterals = 104 × 60 = 6240 lph or Qs = 6.24 m³/h.

For 60 outlets, F = 0.36, L = 150 m, Le = 60 × 0.36 = 21.60 m C = 140, and assuming D = 50 mm, we get: $J = 1.526 \times 10^4 (6.24/140)^{1.852} \times 5^{-4.87} (150 + 21.60) \times 0.36 = 1.17$ m

Head at the inlet of submain = Hsubmain + Hlat + Hemitter + Hslope = 1.17 + 0.26 + 10.00 + 0.00 = 11.43 m

Head loss in the submain = 11.43 − 10.26 = 1.17 m

Variation in head = $\dfrac{1.17}{11.43}$ × 100 = 10.24%, which is less than 20%.

Hence, the size of 50 mm PVC pipe can be accepted for submains.

3.5.1.9   SIZE OF MAIN

Length of main = 50 m

Total discharge of main = discharge of submain = 6.24 m³/h

Now, assuming D = 50 mm, we get:

$J = 1.526 \times 10^4 (6.24/140)^{1.852} \times (5)^{-4.87}(50 + 1) \times 1 = 0.96$ m

Head at the inlet of main = 11.43 + 0.96 = 12.39 m

Variation in head = $\dfrac{0.96}{12.39}$ × 100 = 7.75%, which is 20%, therefore, use 50 mm dia of main.

3.5.1.10   TOTAL DYNAMIC HEAD (TDH)

1. Pressure head at emitter = 10.00 m
2. Head loss in lateral = 0.26 m
3. Head loss in submain = 1.17 m
4. Head loss in main line = 0.96 m
5. Static head = 10 m
6. Head loss due to filters = 7 m
7. Head loss due to fertilizer injector = 1 m
8. Head loss due to main control valve = 0.50 m
9. Head loss due to fittings etc. = 3.09 m
    TDH = sum of all = 33.98 m

3.5.1.11 PUMP SIZE (HP)

$$HP = \frac{QH}{75n} = \frac{\left(\frac{6240}{3600}\right) \times 33.98}{75 \times 0.65} = 1.3 \text{ hp or use 2 hp pump for operating a drip}$$

system for 1.5 ha of pomegranate.

3.5.1.12 DETAILS OF DRIP UNIT

1. Drippers: 1560, 4 lph each, two drippers per tree.
2. Laterals: 60, each 50 m, total length = 3000 m, dia. 12 mm, LLDPE
3. Submain: 1, total length = 150 m, dia. 50 mm, PVC
4. Main: 50 m long, dia. 50 mm, PVC
5. Pump: 2 hp
6. Sand and screen filter coupled with by-pass and back flush system.
7. Fittings and accessories: tees, bends, control, values, grommet and take-off, and flushing caps, cement solution, pressure gages etc.

The cost of the drip irrigation system can be estimated by considering the current market rates.

3.6 DESIGN OF DRIP IRRIGATION SYSTEM FOR BANANA

1. Field size: 100 m x 150 m
2. Soil: Medium textured soil
3. Slope: 0.3% South-north
4. Peak rate of PE: 10 mm/day
5. Source of water: Well
6. Location of well: South-east corner of field
7. Available discharge: 5.6 lps
8. Static head: 9.0 m.
9. Crop variety: G-9 (Grand nine)
10. Spacing: 1.75 × 1.75 m
11. Crop coefficient (kc): 1.0
12. Pan factor (kp): 0.7
13. Irrigation scheduling: Alternate day basis

14. Age   Kc
    0–2 months   0.4
    3–5 months   0.6
    6–8 months   0.8
    9–12 months   1.0
    > 13 months   0.9

3.6.1   DESIGN PROCEDURE

3.6.1.1   NET DEPTH OF WATER REQUIRED (ETC)

PI = 10 mm/day for alternate day = $10 \times 2 = 20$ mm

Etc = PE × kp × kc = $20 \times 0.7 \times 1.00 = 14.00$ mm

If the emission uniformity is 90%, the gross depth = $14/0.90 = 15.55 = dg$

3.6.1.2   VOLUME OF WATER

V = A × Aw × dg = $1.75 \times 1.75 \times 0.50 \times 15.15 = 23.191$/plant on alternate day.

    Where V = Volume of water in liters/plant/alternate day, A = Area/plant (Row spacing × Plant spacing), Aw = Wetted area factor/(0.5), and dg – Net depth required, mm.

3.6.1.3   EMITTER SELECTION AND NUMBER OF EMITTERS PER LATERAL

No. of emitters/Lateral = [Length of lateral in m] ÷ [Plant spacing (m) × No. of emitter/plant] = $50/1.75 \times 2 = 57.14$ or 58.

    The water requirement is 23.19 liters/plant/alternate day. Therefore, use two emitters per plant of 4 lph.

3.6.1.4   TIME OF IRRIGATION

Time of irrigation = [Volume of water to be applied (Liter)] ÷ [Emitter discharge rate, lph] = $23.19/8 = 2.89$ h = 2 and 53 min.

    For two emitter = 8 lph.

3.6.1.5   NUMBER OF LATERALS TO BE OPERATED AT A TIME

Lateral length = 50 m

No. of emitters/Lateral = 58

Discharge of lateral = (No. of emitters/later) × emitter discharge (lph) = 58 × 4 = 232 lph

Discharge available = 5.6 lps or 20,160 lph

No. of laterals operated at a time = [Total discharge available, lph] ÷ [Lateral discharge, lph] = 20,160/232 = 86.9 or 87.
There will be two units for whole field. The half portion of whole field will have 86 (150 m length/1.75 laterals spacing) = 85.71 say 86 = Numbers of laterals of 50 m length. At one time, only one unit will be operated.

3.6.2   LATERAL SIZE ESTIMATION

Assume: Permissible head loss = 20% (in specific condition), and permissible discharge variation = 10%. To find out size of lateral (diameter of lateral), the head loss due to friction must be found. The Hazen–William's equation to determine head loss due to friction in pipe is: $J = 1.526 \times 10^4$ (Q/c) 1.852 x $D^{-4.87} \times (L + Le) - Fpipe$.

In drip lines, the online and inline emitters are connected. They will protrude inside the tube and offer resistance to flow. This extra resistance needs to be taken into account this is expressed in terms of equivalent length.

Type	Le (m)/emitter
Online drippers	0.1–0.6
In-line drippers	0.3–1.00

Using D = 1.6 cm, C = 130, Q_{lat} = 232 lph or 0.232 m³/hr, lateral length = 50 m, f = 0.36, equivalent length (Le) = emission points × outlet factor(f) = 58 × 0.36 = 20.88 m

$J = 1.526 \times 10^4$ (Qc) $^{1.825} \times (1.6)^{-487} \times (L + Le)$ (f)

= 1.526×10^4 (0.232/130) 1.852 × (1.6) −4.87 × (50 + 20.88) × 0.36 = 0.32 m

Permissible head loss variation = 20%

Operating head of emitter = 10 m

Permissible head = 10 × 0.2 = 2 m > 0.32 m. Therefore, design is safe, select lateral diameter of 16 mm.

3.6.2.1   DESIGN OF MANIFOLD

Head at the outlet of manifold = Operating head of emitter + head loss in lateral = 10 + 0.32 m = 10.32 m

No. of manifolds = Two

Length of each manifold = 75 m

Manifold discharge = Lateral (q) × No. of laterals/manifold = 232 × 86 = 19,952 lph = 19.952 m³/h.

Use f = 0.36, Le = emission outlets × f = 86 × 0.36 = 30.96 m, C = 140, and assuming

D = 63 mm = 6.3 cm, we get:

$J = 1.526 × 10^4 × 0.027 × (6.3)^{-4.87} × (75 + 30.96) × 0.36 = 2.01$ m.

Head at inlet of manifold = $He + H_L$ + head loss in manifold (Hm) + head loss or gain due to slope (H_s)

Slope = 0.3% south-north, and the manifolds are laid along the slope; hence, there is gain in pressure.

Head at inlet of manifold = 10 + 0.32 + 2.01 – 0.22 = 12.55 m

Variation in head = 100 × [12.55 – 10.32]/12.55 = 17.8% < 20.99%, hence it is safe. Hence, select 63 mm diameter of manifold.

3.6.2.2   SIZE OF MAIN

Length (main) – 100 m

Q = same as of manifold = 19.952 m³/h,

C = 140, Le = emission points × f = 1 × 1 = 1 m, assuming D = 7.5 cm we get:

$J = 1.526 × 10^4 (19.952/140)^{1.852} × (7.5)^{-4.87} × (100 + 1) × 1 = 2.28$ m

Head at inlet of main = Head loss in main (Hm) + Head loss at inlet of manifold (He) + Head loss in lateral (Hl) + Pressure head at emitter (He) + Head loss on gain due to slope (Hs) = 2.28 + 2.01 + 10 + 0.32 – 0.15 = 14.46 m

Variation in = 100 × [(14.46 – 12.55)]/14.46 = 13.21% <20%. Design is safe.

Select main size of 75 mm or 7.5 cm.

3.6.2.3   TOTAL DYNAMIC HEAD (TDH)

TDH = H at inlet of main + static head + head loss due to fitting (10% of total) + head loss due to filters (3–4 m) = 14.46 + 9.0 + 2.34 + 4 = 29.8 m

3.6.2.4 PUMP SIZE (HP)

HP = [Qpum (lps) × TDH (m)]/[75 × n (pump efficiency)]
Where: n = 65%, Qpump = 19,952 lph = 19,952/3600 lps = 5.54 lps, and
TDH = 29.8 m.
HP = [5.54 × 29.8]/[75 × 0.65] = 3.38 hp
Select pump size of 4 hp.

3.7 DESIGN OF DRIP IRRIGATION SYSTEM FOR CAULIFLOWER

1. Size of land: 80 m × 100 m
2. Type of soil: medium black
3. Land slope: uniform slope
4. Maximum evaporation: 9.5 mm/day
5. Water resource: well suited at north-south corner of field
6. Available discharge: 6.5 lps
7. Static head:10 m
8. Wetted area: 70%
9. Life of crop: newly planted
10. Crop type: Cauliflower (cv. Golden 80) on 0.8 ha.
11. Spacing: 0.60 m × 0.40 m

3.7.1 DESIGN PROCEDURE

3.7.1.1 NET DEPTH OF WATER

Evapotranspiration of crop (ET) = EP × Kp × K_c × 9.5 × 0.75 × 0.8 = 5.70 mm/day

3.7.1.2 VOLUME OF WATER

Total volume of water required = plant spacing × row spacing (m) × wetted area (in fraction) × depth of water (cm) = 0.6 × 0.4 × 0.7 × 5.70 = 0.95 l/day/plant
Use one emitter for three plants. Therefore for three plants:
Water requirement = 3 × 0.95 = 2.85l/day/three plants
Consider 90% emission uniformity of the drip irrigation system.

Volume of water = 2.85/0.90 = 3.16 l/day/three plants or per emitter or = 3.16 l/day/emitter.

3.7.1.3   EMITTER SELECTION AND POSITIONING

To economize the cost of the system, use one lateral for three rows of cauliflower. Therefore take the lateral spacing of 1.2 m and the soil is medium black. The lateral movement of the water will be faster than the vertical movement. Therefore, take emitter spacing of 0.60 m. One emitter will irrigate three plants of cauliflower (as per layout).

3.7.1.4   OPERATION TIME

$$\text{operation time (To)} = \frac{\text{volume of water applied / tree}}{\text{total dripper discharge}} = \frac{3.16}{4} = 0.79\,\text{h} = 47.5\,\text{min.}$$

3.7.1.5   DETAILS OF DRIP UNIT

1. Length of lateral = 40 m.
2. Spacing between two successive emitters.
3. Number of plants per lateral = 40/0.66 = 66.66 for single row. We are putting a lateral for three rows. Therefore total number of plants on one lateral = 200 plants.
4. Number of drippers per lateral = 66.66 or 67.
5. Total no of laterals = 100/120 = 83; one manifold on either side. There will be = 83 + 83 = 166 laterals each measuring 40 m in length.
6. Total number of emitters = There are 67 emitters on one lateral = 166 × 67 = 11,122 emitters.
7. Total length of lateral = 166 × 40 = 6640 m.
8. Discharge through one lateral = 67 × 4 = 268 lph.

3.7.1.6   NUMBER OF SHIFTS

Discharge through one lateral = 67 drippers × 4 lph = 268 lph

$$\text{No of laterals that can be operated at a} = \frac{\text{discharge available per hr}}{\text{disharge through one lateral}} =$$

$$\frac{6.5 \times 3600}{268} = 87.31 \text{ or 87 laterals can be operated at a time.}$$

However, we have to split the whole field in two units and to operate accordingly as per layout.

3.7.1.8 SIZE OF LATERAL

The permissible head loss due to friction by taking into account 10% head variation will be 10 m × 0.10 = 1 m (for 4 lph dripper operated at 10 m head). The Hazen–William equation for head loss due to friction is:

$$J = 1.56 \times 10^4 (Q/C)^{1.852} D^{-4.87} (L + L_e) F$$

Using f = 0.36 for 67 outlets on laterals, L = 40 m, L_e = emission holes × factor = 67 × 0.36 = 24.12 m, C = 130, and assuming D = 16 mm.

$$J = 1.56 \times 10^4 (0.268/130)^{1.852} (1.6)^{-4.87} (40 + 24.12) \times 0.36 = 0.37 \text{ m} < 1 \text{ m}.$$

Lateral of 12 mm size is sufficient to meet the hydraulic requirements.

3.7.1.9 SIZE OF SUBMAIN

Head loss at the outlet of submain = 10 + 0.37 = 0.37 m

Number of submian = 2, length of one submian = 50 m

Discharge (Q) through one submain = Qlat × No. of laterals = 268 × 83 = 22,244 lph or Q_s = 22.244 m³/h

Using: for 83 outlets, F = 0.36, L = 50 m, L_e = 83 × 0.36 = 29.88 m, C = 140, and assuming D = 50 mm: J = 1.56 × 10⁴ (22.244/140)^{1.852} (5.0)^{-4.87} (50 + 29.88) × 0.36 = 5.73 m

Head at the inlet of submain = Hsubmain + Hlat + Hemitter + Hslope = 5.73 + 0.37 + 10.00 + 0.00 = 16.10 m

Head loss in the submain = 16.10 – 10.37 = 5.73 m

Variation in head = 5.73/16.10 × 100 = 35.59% >20%. Hence, the size of 50 mm PVC pipe cannot be accepted for submains. Therefore for 63 mm size of submain = Hsubmain + Hlat + Hemitter + Hslope = 1.86 + 0.37 + 10.00 + 0.00 = 12.23 m

Head loss in the submain = 12.23 – 10.37 = 1.86 m

Variation in head = [1.86/12.23] × 100 = 15.21% <20%. Hence, the size of 63 mm PVC pipe can be accepted for submains. Therefore, take submain of 63 mm dia.

3.7.1.10   SIZE OF MAIN

Length of main = 100 m

Total discharge of main = discharge of submain = 22.244 m³/h.

Assuming D = 75 mm: J = $1.56 \times 10^4 (22.244/140)^{1.852} (7.5)^{-4.87} (100 + 1) \times 1 = 2.79$ m

Head at the inlet of main = 12.23 + 2.79 = 15.02 m

Variation in heads = [2.79/15.02] × 100 = 18.57% <20%. Hence use 75 mm dia of main.

3.7.2   TOTAL DYNAMIC HEAD (TDH)

1. Pressure head at emitter = 10 m
2. Head loss in lateral = 0.37 m
3. Head loss in submain = 1.86 m
4. Head loss in main line = 2.79 m
5. Static head = 10.00 m
6. Head loss due to filters = 7 m
7. Head loss due to fertilizer injection = 1 m
8. Head loss due to main control valve = 0.50 m
9. Head losses due to fittings etc. = 3.35 m
    a. Fittings at the rate of 10% for safety factor
        TDH − sum of all heads = 36.87 m

3.7.2.1   PUMP SIZE (HP)

$$H.P. = \frac{Q.H.}{75n} = \frac{22244/3600}{75 \times 0.65} \times 36.87 = 4.67 \text{ hp or 5 hp}$$

Use pump of 5 hp for running a drip system for 0.8 ha of cauliflower.

3.7.2.2   DETAILS OF DRIP UNIT

1. Drippers: 11,122; 4 lph, one dripper per three plants.
2. Laterals: 166, each 40 m, total length = 6640 m, dia. 12 mm, LLDPE.
3. Submain: 2, total length = 100 m, dia 63 mm, PVC.
4. Main: 100 m length, dia 75 mm, PVC.
5. Pump: 5 hp.

6. Sand and screen filter coupled with by –pass and black flush system.
7. Fittings and accessories: tees, bends, control, values, grommet and take-off, and flushing caps, cement solution, pressure gauges etc.
8. The cost of the drip irrigation system can be estimated by considering the current market rates.

3.8 DESIGN OF DRIP IRRIGATION SYSTEM FOR SUGARCANE

1. Land size: 100 m × 100 m
2. Soil type: Medium black soil
3. Slope: 0.3% (south-north)
4. Max. evaporation/day: 12 mm
5. Water source: Well at South right corner of the field
6. Available discharge: 3.5 lps
7. Static head: 10 m
8. Effective root zone depth: 60 cm
9. Percentage wetted area: 60%
10. Crop: Sugarcane with paired row planting (75–150–30 cm) on 1 ha.

3.8.1 DESIGN PROCEDURE

3.8.1.1 DEPTH OF IRRIGATION

Evapotranspiration = PE × PF × KC = 12 × O.7 × 1.15 = 9.66 mm/day

3.8.1.2 VOLUME OF WATER PER EMITTER PER DAY (V)

V = Spacing between lateral × spacing between emitter × % wetted area × depth *of* water = 2.25 × 0.75 × 0.60 × 9.66 = 9.78 l/day/emitter.
With emission uniformity of 90%, the volume in l/day/emitter is:

$$= \frac{\text{Volume of water applied} \frac{\text{emitter}}{\text{day}}, \text{lit}}{(EU)\text{emission uniformity } \%} = \frac{9.78 \frac{\text{lit}}{\text{day}} / \text{emitter}}{0.90} = 10.86 \text{ or}$$

11 l/day/emitter

3.8.1.3   EMITTER SELECTION

The volume *of* water to be delivered per day per emitter is **11** l. Let us select the emitter of 4 lph discharge.

3.8.1.4   IRRIGATION TIME (T_o)

$$To = \frac{11}{4} = 2.75 \text{ h or 2 h and 45 min.}$$

3.8.1.5   LAYOUT OF DESIGN

1.   Area = 100 m × 100 m = 1 ha
2.   Length of lateral = 50 m
3.   Spacing between laterals = 2.25 m
4.   Spacing between emitter = 0.75 m
5.   No of emitter per lateral = $\frac{50}{0.75}$ = 66.66 or 67 with each emitter of 4 lph.
6.   The discharge through one lateral = Qlateral = 67 × 4 = 268 lph
7.   Number of laterals that can be operated at a time and number of shifts: Available discharge, 3.5 lps = 3.5 × 3600 = 12,600 lph

   No. of laterals that can be operated at time = $\frac{1200}{268} = 47$

   Maximum 47 laterals can be operated at a time. This information is useful in deciding the number of manifolds required for the layout. The width of the field is 100 m and the spacing between two laterals is 2.25 m.

   No. of laterals = $\frac{100}{2.25}$ = 44.44 or 44 on one side.

   On both the sides = 44 × 2 = 88 laterals, each of 50 m in length.

   No of manifolds = $\frac{88}{47}$ = 1.87 or 2.

   No. of laterals/manifold = 44.

   If the laterals are laid across the slope (east-west) in both sides (I) and (II) of field, then all the laterals of any portions will be operated at a time. Therefore, whole drip system can be run in two units (shifts).

3.8.1.6   DESIGN OF LATERAL

The permissible head loss due to friction, by taking into account 10% variation, will be

$10 \times 0.10 = 1.0$ m (4 lph emitter operated at 10 m head).

Applying Hazen–William's equation used for determining the size of lateral:

$$J = 1.56 \times 10^4 \, (Q/C)^{1.852} \, (D)^{-4.87} \, (L + L_e) \times 0.36$$

Using $f = 0.36$ for 67 emission points, $L = 50$ m, $Le = 67 \times 0.36 = 24.12$ m, $C = 130$ and Qlateral = 268 lph = 0.268 m³/h, we get $J = 1.78$ m >1.0 m.

Therefore, the lateral size of 12 mm is not sufficient to meet the hydraulic requirement. Hence, the design must be revised. Now using $D = 16$ mm

$$J = 1.526 \times 10^4 \, (0.268/130)^{1.852} \times (1.2)^{-4.87} \times (50 + 24.12) \, 0.36 = 0.4381$$
< 1 m

Hence, select the lateral size of 16 mm in diameter.

3.8.1.7   DESIGN OF MANIFOLD

Head loss at the outlet of manifold = $10 + 0.44 = 10.44$ m

Qmanifold = Qlateral × No. of laterals on manifold (= 2×44) = 11,792 lph = 11.792 m³/h

Using $f = 0.36$ for 44 outlets, $L = 50$ m, $Le = 44 \times 0.36 = 15.84$, $C = 140$, and assuming $D = 50$ m, we get:

$$J = 1.526 \times 10^4 \, (11.792/140)^{1.852} \times (5.0)^{-4.87} \times (50 + 15.84) \times 0.36 = 1.45 \text{ m}$$

$$Hs = \frac{50 \times 0.3}{100} = 0.15 = \text{head loss due to 0.3\% slope in the south-north}$$

direction.

Now head loss (H) at the inlet of manifold = Hm = Head loss due to main = Htn + H1 + He + HS = $1.45 + 0.44 + 10.0 + 0.15 = 12.049$

Difference of head in manifold = $12.04 - 10.44 = 1.60$ m

variation in head $= \dfrac{1.61}{12.04} \times 100 = 13.30\%$, which is <20% allowable variation in head.

Hence, we use $D = 63$ mm.

3.8.1.8   SIZE OF MAIN

Length of main line = $50 + 50 = 100$ m

Qmain = Qmanifold = 11.792 m³/h
Using C = 140 and D = 50 mm, we get:
J = 1.526 × 10⁴ (11.792/140)^{1.852} × (5.0)^{-4.87} × (100 + 1) × 1 = 6.22 m Now
head at the inlet of main = 12.04 + 6.22 = 18.26 m

variation in head = $\dfrac{6.22}{18.26}$ × 100 = 34.06% > 20%, allowable variation
in head.

Hence, the design cannot be accepted.
Now assume D = 63 mm.
J = 1.526 × 10⁴ (11.792/140)^{1.852} × (6.3)^{-4.87} × (100 + 1) × 1 = 2.01 m
Head at the inlet of mainline = 12.04 + 2.01 = 14.05 m
Differenced of head in mainline = 14.05 − 12.04 = 2.01 m

variation in head = $\dfrac{2}{14.05}$ × 100 = 14.30% < 20%.

Hence, the design is safe. Therefore use D = 63 mm for the mainline.

3.8.1.9   TOTAL DYNAMIC HEAD (TDH)

H_{Loss} = 10% of sum all heads
TDH = Head at inlet of manifold + H main + H static + H_{Loss}
= 12.04 + 2.01 + 10 + 2.40 = 26.45 m

3.8.1.10   PUMP SIZE (HP)

$$HP = \frac{Q \times H}{75 \times n} = \frac{11792 / 3600}{75 \times 0.65} \times 26.45 = 1.78 \text{ or } 2 \text{ hp}$$

Select a pump of 2 hp such that it can give minimum discharge of 3.5 lps
at the total head of 26.45 m.

3.8.1.11   COST ESTIMATION OF DRIP IRRIGATION SYSTEM

1. Cost of emitters
    No. of emitters per lateral = 67
    No. of emitters for all laterals = 67 × 88 (There are 88 laterals on both
    sides of manifold) = 5896

    Total cost of emitters at the rate of Rs. 6 each (pressure compensating) =
5896 × 6 = Rs. 35,376.00

2. Cost of laterals
   No. of laterals required = 88 each of 50 m in length.
   Total length of laterals = 88 × 50 = 4440 m
   Cost of lateral at the rate of Rs 5/m = 4440 x 5 = Rs. 22,000.00
3. Cost of filtration system = Rs. 5000.00
4. Cost of pressure gages = Rs. 500.00
5. Cost of control valve
   Cost of four control valves at the rate of Rs. 150 each = 4 × 150 = Rs. 600.00
6. Cost of main and manifold
   Manifold has a length of 100 m with 50 mm diameter pipe.
   Cost at the rate of Rs. 25 per meter = 100 × 25 = Rs. 2500.00
7. Cost of main
   Main line has a length 100 m with 63 mm diameter pipe.
   Cost at the rate of Rs. 30 per m = 100 × 30 = Rs. 3000.00
8. Electrical motor along with its fitting = Rs. 40,001.00
9. Fitting charges = Rs. 500.00
10. Total cost = sum of items 1–9 = Rs. 64,580.00
    Life of drip irrigation system = 5 years
    Therefore average investment = 54,580/5 = Rs. 12,916 per annum and for 1 ha.

KEYWORDS

- banana
- cauliflower
- clogging
- cost estimation
- design
- drip irrigation
- emission efficiency
- emission point
- emitter
- guava
- Hazen–William

- **head loss**
- **India**
- **Indian conditions**
- **orange**
- **permissible variation**
- **pomegranate**
- **sugarcane**
- **water scarcity**

REFERENCES

1. Goyal, M. R.; Harmsen, E. W. eds., *Evapotranspiration: Principles and Applications for Water Management*. Apple Academic Press Inc.: Oakville, ON, 2014.
2. Goyal, M. R. ed., *Research Advances in Sustainable Micro Irrigation.* Apple Academic Press Inc.: Oakville, ON, 2015; Vol. 1–8.
3. Goyal, M. R. ed., *Innovations and Challenges in Micro Irrigation.* Apple Academic Press Inc.: Oakville, ON, 2016; Vol. 1–4.

CHAPTER 4

GRAVITY-FEED DRIP IRRIGATION FOR AGRICULTURAL CROPS

BERNARD OMODEI[1] and RICHARD KOECH[2]

[1]Measured Irrigation, Dot2dot Post Pty Ltd, 5/50 Harvey Street East, Woodville Park SA 5011, Australia. Email: bomodei@ measuredirrigation.com.au

[2]School of Environmental and Rural Science, University of New England, Armidale NSW 2351, Australia. Tel.: +61- 267735221; Email: rkoech@une.edu.au or richardkoech@hotmail.com.

CONTENTS

ABSTRACT

This chapter describes a method of irrigation called measured irrigation (MI). MI is a gravity-feed irrigation system that directly controls the volume of water emitted from each emitter nozzle in each sector during the irrigation event without the need to control the flow rate or the duration of the irrigation event. MI does not require access to electricity grid power or to an urban water supply. The design of unpowered MI is discussed for small plots of approximately 0.01 ha and larger plots of approximately 1 ha. It is shown how variations in the application rate are controlled by the prevailing weather conditions. Step by step instructions are provided for the establishment of unpowered multi-sector MI. It is demonstrated that MI is highly water- and energy-efficient.

4.1   INTRODUCTION TO MEASURED IRRIGATION

Measured irrigation (MI) is a low pressure micro irrigation system that controls the application rate to each plant. The application rate to each plant is directly proportional to the net evaporation (= evaporation minus rainfall). With MI, the plants to be irrigated are often grouped into sectors (zones) whereby the irrigation of each sector is independent of all the other sectors. For each sector, the emitters should satisfy the MI principle which is defined as follows: "For any two emitters in a sector and at the same pressure, the ratio of the flow rates is independent of the pressure within the operational pressure range for the sector."

For MI, an emitter may be a dripper, a length of micro tube, or a nozzle. The term nozzle refers to a short cylindrical tube or hole for restricting the flow. Pressure compensating drippers should not be used for MI.

MI does not require access to an urban water supply or to electricity grid power, and so there are no ongoing costs for reticulated water or electricity. This makes the system particularly suitable to poorer countries, where access to these facilities is either unreliable or too expensive.

For conventional pressurized irrigation systems, the volume of water delivered to a plant during the irrigation event depends upon the flow rate. But for MI, the volume of water delivered to a plant during the irrigation event is independent of the flow rate. This very important property of MI will be explained in more detail later in this chapter.

The first public presentation of the MI technology was at the 7th Asian ICID regional conference in Adelaide [8]. The following three implementations of MI are discussed in some detail in by other scientists [6, 9,10]: (a) Unpowered MI, (b) Solar-powered single sector MI, and (c) Solar-powered multi-sector MI.

This chapter will focus on low cost unpowered MI for smallholders. The chapter will emphasize the simplicity of MI so that readers may understand the basic principles. With this basic understanding, an interested reader may be able to design a low cost MI system for their particular irrigation requirements using locally available resources and materials. Access to sophisticated manufactured items is not a requirement.

MI had been successfully implemented in a number of community gardens in Australia [5].

4.2 REVIEW OF LOW COST MICRO IRRIGATION SYSTEMS

Micro irrigation has become synonymous with modern and efficient irrigation practices that conserve precious water resources and maximize plant performance [4]. However, the biggest barrier to the adoption of micro irrigation in developing countries has been the installation cost [11].

The two most popular technologies for low cost micro irrigation are gravity-feed drip tube (or drip tape) irrigation and gravity-feed micro tube irrigation. In general, a low cost system needs to be gravity-feed to avoid the additional cost of buying, running, and maintaining a pump. Venot et al. [12] emphasize that the drip irrigation hardware acquires its characteristics only through, and within, the network of institutions, discourses, and practices that enact it. An example of an irrigation kit for each technology is discussed below.

Netafim was awarded the *2013 Stockholm Water Prize* for their contribution to drip irrigation (http://www.netafim.com/Data/Uploads/A4%20data%20sheet%20update%20139.pdf), and in particular for their development and distribution of the *Family Drip System*, a range of gravity-feed dripper-line irrigation kits for smallholders [7]. The Netafim dripper-line technology uses the patented TurboNet™ dripper technology and is more expensive than the simpler micro tube technology, which is in the public domain [2]. On the other hand, for almost 15 years, International Development Enterprises (IDE) have been promoting and distributing their range of gravity-feed micro tube irrigation kits to smallholders in developing countries [3].

The water efficiency of both of these gravity-feed technologies depends upon the smallholder knowing the flow rate from the emitters and controlling the duration of the irrigation event. Knowing the flow rate from the emitter and the duration of the irrigation event, smallholder can calculate the volume of water delivered to each plant during the irrigation event. Both technologies have the following disadvantages which can be significantly reduced by upgrading to MI.

- It is recommended that base of the tank should be at least 1 m higher than the emitters (for MI the base of the tank can be much lower, depending upon the specific application).
- As the water level in the tank falls during the irrigation event, the flow rate at the emitters will also fall. However, the volume calculations assume that the flow rate at the emitters is constant.
- When the land is uneven or sloping, errors will occur in the calculation of the volume of water delivered to each plant. Netafim recommends that the slope of the land should be less than 2% [7].
- Neither technology adjusts the application rate to the plants throughout the year according to the prevailing weather conditions. Reductions in water efficiency are likely to occur because of poor decisions made by the smallholder in relation to variations in the application rate (for MI the application rate for each plant is automatically adjusted to take account of the prevailing weather conditions).

It is appropriate to regard unpowered MI as an extension or refinement of low cost gravity-feed micro irrigation systems.

4.3   UNPOWERED MULTI-SECTOR MI

A schematic diagram for unpowered MI with two sectors is shown in Figure 4.1. This application has been chosen because it is very simple and it demonstrates the basic principles of MI. Furthermore, the application demonstrates the advantages of MI over other more complicated irrigation technologies. The various components of unpowered multi-sector MI (Fig. 4.1) are described in this section.

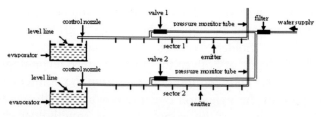

FIGURE 4.1   Schematic diagram of unpowered MI with two sectors.

4.3.1 EVAPORATOR

An evaporator is an open container with vertical sides to ensure that the surface area for evaporation is constant regardless of the volume of water in the evaporator. A level line is marked on the inside of the container about 3 cm below the overflow level. Figure 4.2 illustrates a plastic hobby box, an example of an evaporator where the surface area for evaporation is approximately 0.1 m². In the application shown in Figure 4.1, there are two evaporators, one for each irrigation sector. The evaporators in the various sectors do not need to be the same. However, for the application in Figure 4.1, it is assumed that the evaporators are identical. Each evaporator should be exposed to full sun so that the water in the evaporator can freely evaporate and so that any rain falling directly above the evaporator enters the evaporator.

FIGURE 4.2 Evaporator.

4.3.2 IRRIGATION SECTOR

An irrigation sector for unpowered multi-sector MI is a set of emitters and the associated network of delivery pipes and tubes. A valve is connected to the sector so that all the emitters in the sector may be isolated from the rest of the irrigation system. All the emitters within a sector should be at approximately the same level. The application in Figure 4.1 has two sectors. Valve 1 is connected to sector 1 and valve 2 is connected to sector 2. The emitters in sector 1 are at a higher level than the emitters in sector 2.

4.3.3   PRESSURE MONITOR TUBE

A pressure monitor tube is an open vertical tube connected to the delivery network near an emitter whereby the water level in the tube measures the pressure at the emitter. Each sector has a pressure monitor tube to measure the pressure at the emitter with the anticipated lowest pressure in the sector. Figure 4.3 illustrates an example of a pressure monitor tube using acrylic tubing.

FIGURE 4.3   Pressure monitor tube.

4.3.4   CONTROL NOZZLE

For each sector, one of the emitters in the sector drips water into the evaporator during the irrigation event and this emitter is called the control nozzle. The control nozzles in the various sectors do not need to be the same. However, for the application in Figure 4.1, it is assumed that the control nozzles are identical.

For each sector, when the water level is about 1 cm below the level line, the valve for the sector is opened. When the water level reaches the level line the valve is closed. Due to evaporation, the water level will fall and so the cycle continues indefinitely.

When it is very hot, the water evaporates more quickly and so the valve is opened sooner. And when it rains, extra water enters the evaporator and so the start of the next irrigation event is delayed.

4.3.5 NOZZLE RATIO

For any emitter in any sector, the nozzle ratio is the ratio of the flow rate of the emitter to the flow rate of the control nozzle for the sector, when both are at the same pressure. The principle of MI ensures that the nozzle ratio is independent of the pressure. The nozzle ratio is a characteristic of the emitter and the control nozzle.

Each sector may have many different emitters depending on the requirements of the plants within the sector. A smallholder may make low cost nozzles using locally available materials. A nozzle is simply a device with a small hole in it to restrict the flow to a plant. For each different nozzle type the landholder will need to measure the nozzle ratio.

For any combination of emitter and control nozzle, there is a simple method to work out the nozzle ratio. Over the same period of time collect the water from the emitter in one container and the water from the control nozzle in another container. Then the nozzle ratio is simply the ratio of the water volumes in the two containers. For this calculation, it is very important that the emitter and the control nozzle are at the same pressure. Using this method, it is very easy to make and calibrate nozzles suited to particular irrigation requirements.

As an example only, the nozzles in Table 4.1 and Figure 4.4 are available from the MI website: www.measuredirrigation.com (MI 2015). The nozzle ratios for any combination of emitter nozzle and control nozzle are shown in Table 4.2.

TABLE 4.1 Nozzles Available from the Measured Irrigation Website.

Nozzle #	Nozzle name	Description
N1	MS dripper	Netafim Miniscape (Landline 8) dripper in dripperline
N2	Green	Stainless steel needle nozzle 0.56 mm ID
N3	Yellow	Stainless steel needle nozzle 0.64 mm ID
N4	Brown	Stainless steel needle nozzle 0.79 mm ID
N5	Pink	Stainless steel needle nozzle 0.99 mm ID
N6	White	Stainless steel needle nozzle 1.17 mm ID
N7	Purple	Stainless steel needle nozzle 1.35 mm ID
N8	Orange	Stainless steel needle nozzle 1.51 mm ID
N9	Olive	Stainless steel needle nozzle 1.77 mm ID

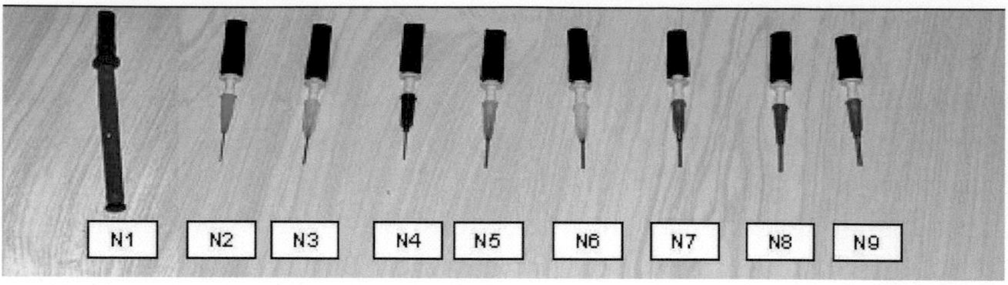

FIGURE 4.4 Nozzles available from the website for measured irrigation.

TABLE 4.2 Nozzle Ratios.

		Control nozzle								
		MS dripper	Green	Yellow	Brown	Pink	White	Purple	Orange	Olive
Emitter nozzle	Nozzle 1 MS dripper	1.00	0.529	0.351	0.207	0.122	0.077	0.0611	0.0436	0.0338
	Nozzle 2 green	2.08	1.00	0.663	0.392	0.231	0.145	0.1155	0.0825	0.0639
	Nozzle 3 yellow	3.13	1.51	1.00	0.591	0.348	0.219	0.174	0.124	0.0964
	Nozzle 4 brown	5.30	2.55	1.69	1.00	0.589	0.371	0.295	0.210	0.163
	Nozzle 5 pink	9.00	4.33	2.87	1.70	1.00	0.630	0.500	0.358	0.277
	Nozzle 6 white	14.3	6.88	4.56	2.70	1.59	1.00	0.794	0.568	0.439
	Nozzle 7 purple	18.0	8.7	5.74	3.40	2.00	1.26	1.00	0.715	0.553
	Nozzle 8 orange	25.2	12.1	8.03	4.75	2.80	1.76	1.40	1.00	0.774
	Nozzle 9 olive	32.5	15.7	10.4	6.14	3.61	2.28	1.81	1.29	1.00

Note: The number of liters of water delivered to a sector during the irrigation event is calculated by multiplying the corresponding nozzle ratio in this table by the control.

4.4 APPLICATION RATES FOR UNPOWERED MI

By maintaining the water level at the level line, the volume of water entering the evaporator must match the volume of water that evaporates, assuming that there is no overflow. Hence, for a whole year:

$$C + R = E \tag{4.1}$$

Where C is the annual volume of water emitted by the control nozzle, R is the annual volume of rainwater entering the evaporator, and E is the annual volume of water that evaporates from the evaporator.

Therefore:

$$C = E - R, \text{ or} \tag{4.2}$$

$$C = (e - r) \times A \tag{4.3}$$

Where e is the annual evaporation, r is the annual rainfall, and A is the cross sectional area of the evaporator.

Provided the evaporator never overflows or runs dry, the Eq 4.3 implies that the annual volume of water emitted by the control nozzle is proportional to the annual net evaporation $= (e - r)$.

4.4.1 NOZZLE FORMULA

The nozzle formula states that for any emitter:

$$\text{measured volume} = \text{control volume} \times \text{nozzle ratio} \tag{4.4}$$

The measured volume is the volume of water emitted by the nozzle. The control volume is the volume of water delivered to the evaporator during the same time interval, and the nozzle ratio is the ratio of the flow rate of the emitter to the flow rate of the control nozzle when both are at the same pressure. All MI volumes are predicted by the nozzle formula. For MI, one can apply the nozzle formula provided that the emitter and the control nozzle are at the same pressure.

4.4.2 AN EXAMPLE FOR ADELAIDE, AUSTRALIA

The average annual evaporation in Adelaide is approximately 1.69 m.

The average annual rainfall in Adelaide is approximately 0.54 m.

The internal cross-sectional area of the evaporator, illustrated in Figure 4.2, is 0.109 m^2.

Hence an approximate value for the annual volume of water emitted by the control nozzle is: $(1.69 - 0.54) \times 0.109 = 125$ liters/year, (52 weeks in a year).

Hence the average volume of water emitted by the control nozzle each week is approximately 125/52 = 2.4 liters.

Suppose one has access to the emitter nozzles available from the MI website and illustrated in Figure 4.4. Assuming that the control nozzle is yellow and that all the nozzles are at the same pressure, then the nozzle formula and the nozzle ratios in the Table 4.2 for the yellow control nozzle can be used to predict the average number of liters per week delivered by each of the nine emitter nozzles. As shown above, the yellow control nozzle delivers an average of 2.4 liters per week. So for each nozzle, the average number of liters per week is estimated in Table 4.3 below by multiplying the nozzle ratio by 2.4. The table also estimates the annual number of liters for each nozzle.

TABLE 4.3   Average Application Rates in Adelaide.

Emitter nozzle	Average number of liters per week	Annual number of liters
MS dripper	0.84	48
Green	1.59	86
Yellow (control nozzle)	2.40	125
Brown	4.06	230
Pink	6.89	359
White	10.94	643
Purple	13.78	828
Orange	19.28	1091
Olive	24.91	1638

Note that these estimates of the average irrigation volumes depend only on the nozzle ratio, the annual evaporation and rainfall, and the cross-sectional area of the evaporator. Provided that the evaporator never overflows or runs dry, the estimates are independent of the control volume and the irrigation frequency.

4.4.3   MONTHLY ESTIMATES OF IRRIGATION VOLUMES

It is assumed that at the end of each month the water level in the evaporator is reset to the level line by either running the irrigation or removing water from the evaporator. Provided that the evaporator never overflows or runs dry, then we get:

$$C_m = \max(0, (e_m - r_m) \times A), \text{ for } m = 1, 2, \dots, 12 \qquad (4.5)$$

Where C_m is the volume of water emitted by the control nozzle for month m, e_m is the monthly evaporation for month m, r_m is the monthly rainfall for month m, and A is the internal cross-sectional area of the evaporator.

Eq 4.5 implies that provided the evaporator never overflows or runs dry, the monthly volume of water emitted by the control nozzle is zero for those months when the rainfall exceeds the evaporation and is proportional to the monthly net evaporation $(e_m - r_m)$ for the other months. Using the nozzle formula for any emitter at the same pressure as the control nozzle, the volume of water emitted by the emitter for month m (V_m) is calculated as below:

$$V_m = r \times \max(0, (e_m - r_m) \times A), \text{ for } m = 1, 2, \dots, 12 \qquad (4.6)$$

where V_m is volume of water emitted by the emitter for month m, and r is the nozzle ratio of the emitter to the control nozzle.

From Eq 4.6, it can be observed that the month by month irrigation volumes depend only on the nozzle ratio, the monthly evaporation and rainfall, and the cross-sectional area of the evaporator, and are independent of the control volume and irrigation frequency.

Monthly data for evaporation and rainfall in Australia are available from the *Bureau of Meteorology*. Provided one has access to historical data for the monthly evaporation and rainfall for the locality desired, the above formula (4.6) can be used to estimate monthly irrigation volumes for any emitter at the same pressure as the control nozzle.

4.4.4 WEEKLY ESTIMATES OF IRRIGATION VOLUMES

Weekly estimates of irrigation volumes are based on the monthly estimates. The following formula is used to calculate the weekly estimates throughout the year:

$$W_m = [\{r \times \max(0, (\hat{e}_m - \hat{r}_m) \times A) \times 7\}/N_m], \text{ for } m = 1, 2, \dots, 12 \qquad (4.7)$$

Where W_m is an estimate of the number of liters per week emitted by an emitter in month m, $\hat{e}_m$ is an estimate of the evaporation in month m, $\hat{r}_m$ is an estimate of the rainfall in month m, A is the internal cross-sectional area of the evaporator, and N_m is the number of days in month m.

4.5 LOW-COST MI IN DEVELOPING COUNTRIES

As stated previously, MI does not require access to an urban water supply or to electricity grid power. Therefore, there are no ongoing costs for reticulated water or electricity. MI is a simple and appropriate technology for smallholders in developing countries. The biggest obstacle for a landholder is likely to be the initial capital cost of installing MI. With MI, each plant receives the desired amount of water per week, no more and no less. Furthermore, the irrigation frequency responds appropriately to the prevailing weather conditions.

4.5.1 EXAMPLE

An example is presented to demonstrate use of low cost MI in a developing country. Let us use dripper-line from the Netafim Family Drip System [7] on 100 m² level land (10 × 10 m²).

The schematic diagram in Figure 4.5 shows a typical configuration of unpowered MI for a 10 × 10 m² block on level ground. Because the ground is level, a single sector is used. Month by month estimates of the application rate for each dripper are obtained with Eq 4.7 using the average month by month evaporation and rainfall in Adelaide. The specifications for the irrigation system are as follows:

- Total area irrigated = 100 m², 10 x 10 m²
- Length of 25 mm main = 20 m
- Length of 19 mm sub-main = 20 m
- Length of *Netafim Family Drip System* dripper-line = 200 m
- Spacing between drippers = 0.3 m
- Spacing between rows of dripper-line = 0.5 m
- Total number of drippers = 660
- Internal cross-sectional area of evaporator = 0.109 m²
- Control nozzle is the same as the drippers in the dripper-line
- Application rate in January: 6.58 L/wk per dripper, total 4340 L/wk
- Application rate in February: 6.02 L/wk per dripper, total 3970 L/wk
- Application rate in March: 4.46 L/wk per dripper, total 2940 L/wk
- Application rate in April: 2.44 L/wk per dripper, total 1610 L/wk
- Application rate in May: 0.70 L/wk per dripper, total 460 L/wk
- Application rate in June: 0 L/wk per dripper, total 0 L/wk
- Application rate in July: 0 L/wk per dripper, total 0 L/wk

- Application rate in August: 0.85 L/wk per dripper, total 560 L/wk
- Application rate in September: 1.92 L/wk per dripper, total 1270 L/wk
- Application rate in October: 3.41 L/wk per dripper, total 2250 L/wk
- Application rate in November: 4.81 L/wk per dripper, total 3170 L/wk
- Application rate in December: 5.75 L/wk per dripper, total 3800 L/wk
- Annual irrigation volume: 161 L per dripper, total 106 kL

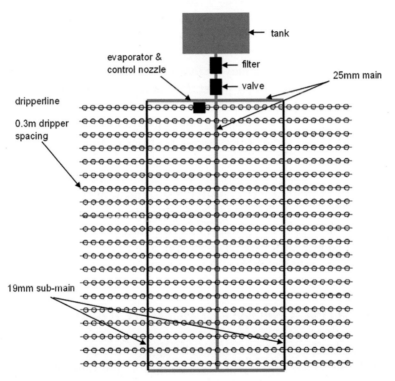

FIGURE 4.5 Schematic diagram of unpowered MI of 10×10 m² on level ground.

Note that the estimates of the application rate assume that the net evaporation (evaporation minus rainfall) for each month is the same as the historical average for the month. The actual application rate is controlled by the prevailing weather conditions.

Power is required if the outlet on the water tank is lower than the drippers. Provided that the water level is less than a meter lower than the drippers, a 20 W solar panel can provide all the power required. For this application, a suitable low pressure pump is a 14 W pump (Model SP20/20) available online from: <www.solarproject.co.uk>.

4.6   MI SYSTEM FOR AREA > ONE HECTARE

Provided that the issue of head loss is managed appropriately, MI can be used for larger blocks of land, where the area is greater than a hectare. To deliver the flows required, a high-flow pump may be required at the reservoir or water tank and so power is required for the pump. Because the final delivery to the plants is by gravity-feed, the pump does not need to generate high pressure. Consequently, the cost and power of the pump will be a fraction of the cost and power of a pump for a conventional pressurized irrigation system.

Depending on the cost, it may be feasible to generate sufficient power for an electric pump using a solar panel. If it is decided that an electric pump is appropriate then the power can also be used to completely automate the irrigation system. More information on solar-powered single-sector MI and solar-powered multi-sector MI is available from Omodei [10] or from the MI Manual [6].

A number of examples using areas of one hectare or more have been successfully modelled using EPANET 2.0 simulation (see the MI Manual [6], Section 17), [1]. The next stage in the development of MI for larger areas is to conduct appropriate field trials. Any organization that has the resources required to conduct such field trials should contact *Measured Irrigation* via the measured irrigation website: <www.measuredirrigation.com>.

4.7   INSTALLING UNPOWERED MULTI-SECTOR MI

This process is illustrated for a particular application in the YouTube video entitled "Think twice before you buy a pump for your rainwater tank":

https://www.youtube.com/watch?v=oN53adj_3sk

The installation instructions below are quite general since the additional details depend upon the choice of water supply, evaporator, emitters, and pipes.

Step 1.   Attach a suitable filter to the water supply. If the water supply is a tank then the base of the tank must be higher than any plant to be irrigated.

Step 2.   Group the plants to be irrigated into sectors. The plants in each sector should all be at approximately the same level.

Step 3.   Select a suitable control nozzle and a suitable range of emitters. Emitters may be unregulated online drippers or unregulated inline drippers in drip tube or tape (do not use pressure compensating drippers). Emitters may also be short lengths of micro tube with varying internal diameter. The dripper-line used for the *Netafim Family Drip System* is ideal.

Step 4.   Determine the nozzle ratio for each type of emitter by using the method described in this chapter.

Step 5.   Estimate the month by month evaporation and rainfall in your locality. This information may be available from the Department of Meteorology.

Step 6.   Select a suitable evaporator and mark a level line on the inside of the evaporator about 3 cm below the overflow level. Measure the surface area for evaporation (this is important because it affects the application rate for the emitter emitters in Step 8).

Step 7.   Set up a spreadsheet based on Eq 4.7 with month by month estimates of the number of liters per week emitted by each type of emitter nozzle.

Step 8   For each plant estimate the number of liters per week required during the hottest dry month of the year. Use the spreadsheet in Step 7 to select the corresponding emitters. For some plants, it may be preferable to estimate the number of liters per week per square meter during the hottest dry month of the year and to use the spreadsheet to determine the appropriate drip tube (or drip tape) spacing. Note that MI does not help the user to decide how many liters per week a plant requires. However, once the user has made a decision, the spreadsheet enables to accurately implement the decision.

Step 9.   Using the emitters selected in Step 8, set up a network of pipes to deliver water from the water supply to the emitters. The emitters in each sector should all be at approximately the same level. Because all emitters within a sector should be at approximately the same pressure, one needs to be aware of any frictional head loss that may occur within a sector and to adjust the design of the network accordingly. Frictional head loss is discussed further in Step 13.

Step 10. For each sector, install a valve so that the sector can be isolated.

Step 11. For each sector install an evaporator in a suitable location exposed to full sun. Install the control nozzle so that it drips water into the evaporator and is at the same level at the emitters in the sector.

Step 12. For each sector install a pressure monitor tube to measure the pressure at the emitter with the anticipated lowest pressure in the sector. Open the valve for the sector and adjust the valve to that the pressure monitor tube shows a suitable pressure for the sector.

Step 13. If it is a concern that the pressure may not be approximately the same at all the emitters within a sector due to frictional head loss, then temporarily connect another pressure monitor tube to measure the pressure near the valve for the sector. If the difference in pressure is not acceptable, then the sector should be redesigned accordingly (see below). Note that a 10% difference in pressure corresponds to a 5% difference in flow rate at identical emitters (in general, a *2x%* difference in pressure corresponds to less than an *x%* difference in flow rate).

4.7.1   REDESIGNING A SECTOR TO REDUCE FRICTIONAL HEAD LOSS

A combination of one or more of the following strategies will reduce frictional head loss.

Strategy 1. If the valve for the sector is connected to a length of pipe that delivers water to many emitters, make sure that the valve is connected to the pipe via a tee junction so the flow rate to the emitters on one arm of the tee balances the flow rate to the emitters on the other arm of the tee.

Strategy 2. Replace the pipe within the sector with pipe with a larger internal diameter.

Strategy 3. Redesign the network of pipes for the sector so that additional piping is used to connect the valve to various locations within the sector.

Strategy 4. As a last resort, one can divide the sector in to two sectors each with their own evaporator and valve.

4.8   CONCLUSIONS

The development of MI as an extension of current micro irrigation systems is at a very early stage. MI is a different conceptual approach to micro irrigation, and each application may require a different implementation of the principle of MI:

For any two emitters in a sector and at the same pressure, the ratio of the flow rates is independent of the pressure within the operational pressure range for the sector.

Below are some of the benefits of upgrading a gravity-feed micro irrigation system to MI:

- Save more water by controlling the application rate for each plant during the hottest dry month of the year.
- Save more water by allowing the prevailing weather conditions to control the variations in the application rate for each plant throughout the year.
- Save more water by maintaining the same level of control of the application rate for each plant throughout the year for sloping ground, regardless of the steepness of the slope.
- The minimum operating head for gravity-feed before upgrade is constrained by the head needed to maintain adequate uniformity of dripper flows. Depending on the particular application, the minimum operating head after upgrade can be lowered without compromising uniformity by grouping the drippers into sectors according to elevation.
- The smallholder can control the irrigation frequency for each sector. For example, if a sector requires more frequent irrigation with less water, then the small-older should open the valve for the sector when the water level is less than 1 cm below the level line.
- The cost of the upgrade on level ground is the cost of a bucket.

KEYWORDS

- **application rate**
- **control nozzle**
- **developing countries**
- **drip irrigation**
- **drip tape**
- **drip tube**
- **dripper**

- **emitter**
- **evaporation**
- **evaporator**
- **flow rate**
- **gravity-feed**
- **irrigation kit**
- **irrigation volume**
- **low pressure**
- **measured irrigation**
- **micro irrigation**
- **nozzle**
- **nozzle ratio**
- **rainfall**
- **smallholder**
- **unpowered**
- **water efficiency**

REFERENCES

1. EPANET. Software that Models the Hydraulic and Water Quality Behaviour of Water Distribution Piping Systems. 2015. Available Online at: http://www.water-simulation.com/wsp/2004/12/22/epanet-2/ (accessed June 1, 2016).
2. Melvyn Kay, Food and Agriculture Organisation of the United Nations. Smallholder Irrigation Technology: Prospects for Sub-Saharan Africa. International Programme for Technology and Research in Irrigation and Drainage, Knowledge Synthesis Report No. 3, FAO: Rome, 2001.
3. IDE. Technical Manual for Ideal Micro Irrigation Systems. IDE & CGIAR Challenge Program for Water and Food. 2011. Available Online at: http://www.ideorg.org/Ourtechnologies/Dripirrigation.Aspx# (accessed April 30, 2015).
4. Lamm, F. R.; Ayars, J. E. *Micro Irrigation for Crop Production;* Nakayama, F. S., Eds.; Elsevier: Amsterdam, Netherlands, 2007.
5. Merry, J.; Langan, S. *Review Paper on 'Garden Kits' in Africa: Lessons Learned and the Potential of Improved Water Management. IWMI Working Paper 162;* International Water Management Institute: Colombo, Sri Lanka, 2014; pp 162.
6. MI. *Measured Irrigation Manual.* 2015. Available Online at: www.measuredirrigation.com (accessed April 30, 2015).

7. NETAFIM. *Family Drip System*. 2015. Available Online at: www.netafim.com/Product/Family-Drip-System (accessed April 30, 2015).
8. Omodei, B. *Measured Irrigation*, 7th Asian ICID Regional Conference: Adelaide, Australia, 2012.
9. Omodei, B. Measured Irrigation: A Significant Development in Water Efficient Irrigation. Transactions of the Wessex Institute of Technology. *Ecol. Environ.* **2013,** *171,* 49–58.
10. Omodei, B. Accuracy and Uniformity of a Gravity-Feed Method of Irrigation. *Irrig. Sci.* **2015,** *33,* 121–130.
11. Postel, S.; Polak, P.; Gonzales, F.; Keller, J. Drip Irrigation for Small Farmers. *Water Int.* **2001,** *26,* 3–13.
12. Venot, J. P.; Zwarteveen, C.; Kuper, M.; Boesveld, H.; Bossenbroek, L.; Van Der Kooij, S.; Wanvoecke, J.; Benouniche, M.; Errahj, M.; De Fraiture, C.; Verma, S. Beyond the Promise of Technology: A Review of the Discourses and Actors Who Make Drip Irrigation. *Irrig. Drain.* **2014,** *63,* 186–194.

PART II
Potential of Solar Energy in Micro Irrigation

CHAPTER 5

PORTABLE SOLAR PHOTOVOLTAIC POWERED PUMPING SYSTEM FOR MICRO-IRRIGATION SYSTEM IN VEGETABLE CULTIVATION

M. K. GHOSAL

Department of Farm Machinery and Power, College of Agricultural Engineering and Technology, Orissa University of Agriculture and Technology, Bhubaneswar 751003, Odisha, India

E-mail: mkghosal1@rdiffmail.com

CONTENTS

ABSTRACT

The growing demands of energy and water particularly in the present agricultural sector have necessitated the adoption of reliable, environment-friendly, and water saving technologies so as to combat against the energy crisis and water stress in near future. It has been established that conventional sources of energy like oil, gas, coal etc. will not be able to provide the desired levels of energy security to humankind in foreseeable future. Hence, there is a global consensus for exploitation and utilization of different renewable energy resources. The search for new options should be eco-friendly as well as abundant in nature. Among the different available renewable energy resources, solar energy seems to be more promising and sustainable.

Solar powered agricultural irrigation may be an attractive application for renewable energy in replacing fossil fuel-powered irrigation devices to achieve energy security. The use of solar photovoltaic (SPV) systems may not only provide good solution for all energy-related problems of the present society but also can perform excellently in terms of productivity, reliability, sustainability, and environmental protection ability. SPV water pumping systems can provide water for irrigation without the need for any kind of fuel or the extensive maintenance as required by diesel and electric pump sets.

Therefore, an attempt was made to develop an affordable and portable SPV water pumping system for irrigating vegetable crops in the state of Odisha. Micro-irrigation method through sprinkler system was integrated with the SPV device to achieve judicious utilization of water. Monthly income of Rs. 15,000 throughout the year was possible by adopting remunerative tomato cultivation on 1 acre of land both during *rabi* and summer seasons in the year 2013 in coastal region of Odisha. Pay-back period of the developed set up was ½ year, due to which it may be easily accepted by the small and marginal farmers of the state in spite of its high initial cost. The popularization of this technology would not only achieve assured water availability to the crops with improved WUE measures by micro-sprinkler irrigation system compared to traditional flood irrigation, but will also protect the environment against release of greenhouse gases and noise pollution by the use of rising diesel and electric pump sets in the state.

5.1   INTRODUCTION

Energy demand is growing exponentially in each segment of the national developments due to the continuous growth and expansion in different sectors like industry, agriculture, irrigation, transportation, communication, housing, health, education, city modernization, entertainment etc. To meet the increasing demands of energy, the share of coal-based power plants for power generation in India is also rising day by day causing severe environmental hazards and thus global warming by releasing a considerable amount of greenhouse gases to the atmosphere. The only alternative in this context is to supplement the existing power sector with non-conventional energy sources. Among the non-conventional energy sources, the solar energy appears to be an attractive and viable proposition because of the abundant and free availability of sun shine in the tropical areas. Moreover, electricity from solar photovoltaic (SPV) system is now gaining more importance because of rapid decline in the cost of SPV modules through advances in research and development in this area. The attention of planners, policy makers, and researchers is also now diverted to the applications of SPV system for pumping of water in irrigation sector due to recent increased water demands in agricultural sector and availability of water has become more crucial than ever before.

In India, electrical and diesel-powered water pumping systems are most widely used for irrigation systems. A source of energy to pump water is also a big problem in developing countries like India. Developing a grid system is often too expensive because rural villages are frequently located too far away from existing grid lines. Even if fuel is available within the country, transporting that fuel to remote and rural villages can be difficult. There are no roads or supporting infrastructure in many remote villages. The use of renewable energy is therefore of utmost importance for water pumping applications in remote areas of many developing countries. Transportation of renewable energy systems, such as photovoltaic (PV) pumps, is much easier than the other types because they can be transported in pieces and reassembled on site [9].

PV energy production is recognized as an important part of the future energy generation because it is non-polluting, free in its availability, and is of high reliability. These facts make the PV energy resource more attractive for many applications, especially in rural and remote areas of the developing countries. Solar PV water pumping has been recognized as suitable for grid-isolated rural locations in places where there are high levels of solar radiation. The state Odisha also receives a good amount of solar radiation for about 4–5 h in a day over a period of nearly 300 days in a year [4]. Solar PV water pumping systems can provide water for irrigation without the need for any kind of fuel or the extensive maintenance as required by diesel and electric pump sets. They are easy to install and operate highly reliable, durable, and modular, which enable future expansion. They can be installed at the site of use, avoiding the spread of long pipelines and infrastructures [1].

Odisha is blessed with highly fertile soil due to flowing of many rivers through it namely the Mahanadi, the Baitarani, the Brahmani, the Subar-narekha, the Budhabalanga, the Bansadhara etc. [3]. As Odisha receives an average annual rainfall of 1500 mm, there is no dearth of water resources. Farmers of the state grow different vegetable crops round the year using hand pump, electric, and diesel pumps for lifting of irrigation water. Lifting of water by hand pump is a most tedious and labor-consuming operation. Similarly, non-availability and erratic supply of grid connected electric supply in the remote areas and rising cost of diesel day by day necessitate the search of a sustainable source of power for assured irrigation particularly for vegetable cultivation, which is nowadays, more remunerative and profitable. Cost of lifting water in the above pumping systems is many folds compared to lifting water by SPV water pumping system [8]. Development of an affordable, durable, and with a very little repair and maintenance would be preferred by the small and marginal farmers of Odisha and India in particular. Installation of electric pump sets is not at all possible at most of the locations as the agricultural fields are far away from the electric grid station. In addition, the electric tariff is increasing every year and thus increasing the cost of water pumping operation. Further, the repair and maintenance cost of electric motor operated pump set is generally more than that of SPV water pumping system [14]. When not much research work was conducted on SPV water pumping system, then, diesel pumping system was very popular among the farming community due to its low cost and portability. During this time, the diesel cost was also cheaper. But it caused a lot of environmental pollution and global warming by the emission of substantial quantity of CO_2 into the atmosphere. The repair and maintenance cost of diesel pump set is also more than that of SPV water pumping system.

Hence, SPV water pumping system is today a viable option left for the farming community as its pumping cost is cheaper compared to electric and diesel pump sets. Moreover, the risk of environmental pollution is less and its repair and maintenance cost is very low. It can be installed at any location as per the desire of the farmers as solar energy is available profusely and free of cost. Portable model of SPV water pumping system would be an added advantage for the farmers looking into the space requirement for permanent installation and fear of theft. Similarly, the need for the optimum utilization of water and energy resources has become a vital issue during the last decade and will become more essential in the future. Hence, the use of solar powered micro-irrigation system is also the need of the hour looking into the present day's concerns of energy crisis and water scarcity particularly in agricultural sector [6].

Therefore, an attempt has been made to develop an affordable and portable solar water pumping system along with micro-irrigation device for irrigating vegetable crops in the coastal regions of Odisha and to study its feasibility among the farmers of the state for growing vegetable crops in their fields for strengthening their livelihoods and socio-economic status. The specific objectives for this research study were thus as follows:

a. Developing a portable and affordable SPV-powered water-pumping device integrated with micro-sprinkler system for irrigating vegetable crops.
b. Feasibility and performance study of SPV-powered micro-sprinkler irrigation system in off-grid remote areas of Odisha.
c. Techno-economic analysis of SPV-powered water-pumping device.

5.2   REVIEW OF SOME DEVELOPMENTS IN SPV WATER PUMPING SYSTEM

The approaches for energy security through SPV power system and improved water use efficiency (WUE) measures through micro-sprinkler irrigation system compared to traditional flood irrigation have thus been thought up nowadays among the researchers, scientists, and agriculturists not only for achieving assured water availability to the crops, but also protecting the environment against release of greenhouse gases and noise pollution by the use of rising diesel pump sets. Attempts have already been made by some researchers in assessing the viability of SPV water pumping system for domestic drinking water and irrigation purposes.

Hamidat et al. [7] studied on small-scale irrigation with PV water pumping system in Sahara regions. The authors developed a mathematical program to test the performance of PV arrays under Saharan climatic conditions. Their study showed that it is possible to use a PV water pumping system for low heads for small scale-irrigation of crops in Algerian Sahara regions. Thus, the PV water pumping system can easily cover the daily water needs of small-scale irrigation with an area smaller than 2 ha. They also concluded that the photovoltaic water pumping system (PVWPS) can improve the living conditions of the farmer with the development of local farming and thus the migration rural workforce will be minimum.

Meah et al. [12] studied on solar photovoltaic pumping system (SPVPS) for remote locations in rural western USA. They realized that solar PVWPS (SPVWPS) is a cost-effective and environmental friendly way to pump water in remote locations, where 24 h electrical service is not necessary and maintenance is an issue. From their survey, it was indicated that a total of 88 SPVWPS were installed in all 23 states of the USA, of which 75 systems were in operation till 2005. They have observed that drought-affected areas like Wyoming, Montana, Idaho, Washington, Oregon, and parts of Texas can use SPVWPS to improve the water supply to livestock in remote locations. They were convinced that successful demonstration of these systems is encouraging other ranchers to try this relatively new technology as another viable water supply option. They concluded that SPVWPS had excellent performance in terms of productivity, reliability, and cost effectiveness, and the system was able to reduce the CO_2 emission considerably over its 25-year life span.

Meah et al. [11] studied on SPV water pumping opportunities and challenges in USA. According to their views, they stated that some improvements can be done to lower the capital investment cost and to reduce the cost of operation and maintenance services using local level operation and maintenance. The authors demonstrated that by using local resources, such as skills, materials, and finances, the SPVWPS could be economically viable in developing countries and competitive with the conventional diesel generator water pumping systems. They concluded that the SPVWPS should be compatible with the local culture and practices to satisfy local wishes and needs, which also could be achieved by using local resources.

Kelley et al. [8] studied the feasibility of solar-powered irrigation in the USA. They developed a method for determining the technical and economic feasibility of PV power irrigation systems applicable to any geographic location and crop type in USA and applied it to several example cases. According to the opinion of authors, the results of technical feasibility analysis agreed with the results obtained from past studies and also showed that there is no technological barrier to the implementation of PVP irrigation if land is available for installation of solar panels. The results of economic feasibility study suggested that the price of diesel has increased sufficiently within the last 10 years to make PVP irrigation economically feasible, despite the high initial cost of PV systems. The authors concluded that as the price of the solar panels is decreasing, the capital costs will decrease making PVP systems even more economically attractive.

Gopal et al. [5] reviewed the research developments on renewable energy source water pumping systems referring 168 research papers across the globe. They concluded that renewable energy solar water pumping systems are identified as an alternative source for replacing conventional pumping methods. The integration of renewable energy sources with water pumping systems plays a major role in reducing the consumption of conventional energy sources and their environmental impacts, particularly for irrigation applications. The SPVWPS are the most widely used renewable energy solar water pumping systems for irrigation and domestic applications, followed by wind energy water pumping systems. The solar thermal and biomass water pumping systems are less popular due to their low thermal energy conversion efficiencies.

Narela et al. [13] studied the feasibility of SPVWPS for irrigating banana plants. They presented the design and economic analysis of efficient SPVWPS for irrigation of banana. The system was designed and installed in solar farm of Jain Irrigation System Limited (JISL), at Jalgaon (Maharashtra). The study area falls at 21° 05′ N latitude, 75° 40′ E longitude, and at an altitude of 209 m above mean sea level. The PV system sizing was made in such a way that it was capable of irrigating 0.165 ha of banana plot with a daily water requirement of 9.72 m^3/day at a total head of 26 m. Also, the life cycle cost (LCC) analysis was conducted to assess the economic viability of the system. The results of the study encouraged the use of the PV systems for water pumping application to irrigate orchards. The installed system of SPVWPS was capable of irrigating 0.165 ha of banana crop within 6.02 h with a daily water requirement of 9.72 m^3/day.

5.3   FUNDAMENTALS OF SPV MICRO-SPRINKLER IRRIGATION SYSTEM

5.3.1   BRIEF HISTORY OF TECHNOLOGY

The PV effect was first discovered by a French scientist Becquerel in 1839 who found that more current could be generated if more light is allowed to fall in the cell. He also discovered that the increase in current is dependent on the wavelength of light. In 1877, Adam and Day first observed the same effect in solids while working with selenium. In 1877, Heinrich Hertz discovered that ultra violet light altered the lowest voltage capable of causing a spark to jump between two metal electrodes. In 1905, Albert Einstein explained that light behaves like a particle rather than a wave. The energy of each light particle, called photon, depends on its frequency only and is equal to the product of Planck's constant (h) and frequency of light (f).

In 1954, researchers at RCA and Bell Laboratories, USA reported achieving efficiencies of about 6% by using devices made of p and n type semiconductors. PV cell most commonly made of silicon, a material called semiconductor, has now been widely used for generating DC electricity.

Historically, it has been about 65 years since the first operational silicon solar cell was demonstrated in 1950s. However, the last 15 years have seen large improvements in the solar cell technology, which is mainly based on the use of silicon. Over the years, solar cell efficiencies have been improved and the cost of solar cell production has decreased. The efficiency of solar cells available in the market is in the range of 13–16%. Silicon-based solar cell modules are now available in the commercial market. There are SPV modules that are made of other technologies, called thin-film technology, like thin-film amorphous silicon, thin-film cadmium telluride etc.

5.3.2   APPLICATIONS OF SPV SYSTEMS

1. Solar street lighting system.
2. Home lighting systems.
3. Water pumping systems (for micro-irrigation and drinking water supply).
4. Space vehicles and satellites.
5. Community radio and television sets.
6. Battery charging.
7. Weather monitoring.
8. Power source for navigational lights.
9. Power source for telecommunication equipment.
10. Power source for railway signaling equipment.

5.3.3   ADVANTAGES OF SPV SYSTEM

1. Absence of moving parts.
2. Direct conversion of light to electricity at room temperature.
3. Can function unattended for long time.
4. Low maintenance cost.
5. No environmental pollution.
6. Very long life.
7. Highly reliable.
8. Solar energy is free and no fuel required.
9. Can be started easily as no starting time is involved.

10. These have high power-to-weight ratio, therefore very useful for space application.
11. Decentralized or dispersed power generation at the point of power consumption can save power transmission and distribution costs.
12. These can be used with or without sun tracking.

5.3.4 LIMITATIONS OF SPV SYSTEM

1. Manufacture of silicon crystals is labor and energy intensive.
2. Low efficiency.
3. The insolation is unreliable and therefore storage batteries are needed.
4. Solar power plants require very large land areas.
5. Electrical generation cost is comparatively higher.
6. The energy spent in the manufacture of solar cells is very high.
7. High initial cost.

5.3.5 AMOUNT OF POWER GENERATED

The amount of power a solar cell can produce depends on solar cell efficiency and solar cell area.

- A 15% efficient solar cell will convert 15% of light falling on to it into electricity.
- Larger the solar cell area, larger will be the power output because in case of larger solar cell, we are collecting light from larger area (a 10 cm by 10 cm) solar cell will have 100 cm^2 (0.01 m^2) area,

Typically, solar cells are characterized by 1000 W/m^2 solar radiation falling on the solar cell at 25 °C solar cell, which is a worldwide standard. Thus, a 0.01 m^2 solar cell, with 15% efficiency under 1000 W/m^2 solar condition will give an output power of:

Power = 15/100% (cell efficiency) × 0.01 m^2 (cell area) × 1000 W/m^2 (Solar radiation) = 1.5 W.

5.3.6 SOLAR CELL CHARACTERISTICS

The schematic diagram of a SPV system is shown in Figure 5.1. The flow of current can be given by diode current equation (Schottky equation):

$$I = I_0 \left[e^{\frac{v}{v_T}} - 1 \right]$$

(5.1)

where, I_o is the reverse saturation current and V_T is the voltage equivalent of temperature = k I/q, k is the Boltzmann constant, T is temperature in Kelvin and q is the charge of an electron.

The current–voltage characteristics of a p–n junction (solar cell) gets modified due to photon or solar generated current (I_{sc}) flowing through the p–n junction as this (I_{sc}) is added with the reverse leakage current (I_o).The diode current equation is now modified as:

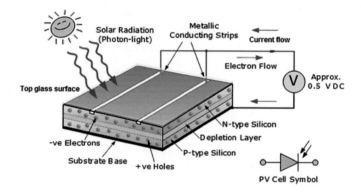

PV Module Anatomy

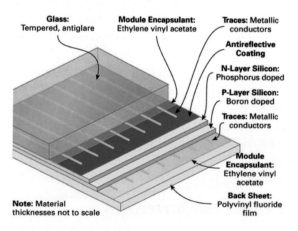

FIGURE 5.1   Construction of a solar cell.

$$I = -I_{sc} + I_0 \left[e^{\frac{v}{v_T}} - 1 \right] \tag{5.2}$$

$$I_{sc} = I_0 \left[e^{\frac{v}{v_T}} - 1 \right] \quad \text{or} \quad V = V_{oc} = V_T \, log \left(\frac{I_{sc}}{I_0} + 1 \right) \tag{5.3}$$

For $V = 0$ (junction is short circuited), $I = 0$, we get Eq. (5.2). V_{sc} is the open circuit voltage.

The above relationship shows that when junction is related with sun's ray and it is short circuited at its terminals, there is a finite current called short circuit current (I_{sc}) that flows through the external circuit made with the short circuiting of the junction terminals. The magnitude of I_{sc} depends upon solar radiation. In case, we use standard convention in which current flowing out from a positive terminal of any energy source, then it is always taken as positive and apply the same convention on a solar cell. The current and voltage characteristic can be redrawn with suitable modification, and mathematically the current–voltage relationship can be written as follows:

$$I = I_{sc} - I_0 \left[e^{\frac{v}{v_T}} - 1 \right] \tag{5.4}$$

The output power from solar cell is the product of voltage and current ($P = I \times V$). It is desirable to operate the solar cell to produce maximum power. Power curve is hyperbola. In case hyperbola of power is drawn on I–V characteristic curve, the hyperbola of power curve is tangential to I–V characteristic at the point of maximum power. Hence, there is only one point on the voltage–current characteristic curve of p–n junction at which the p–n junction produces maximum power for a given insolation or illumination level. In case, we operate the p–n junction at any other point on I–V characteristic curve, power produced will be lesser than as maximum power, resulting in certain amount of solar radiation energy being wasted out at thermal power. The maximum power output can be determined when the product of voltage and current is maximum. The product of voltage and current has the greatest value when the rectangle having sides equal to these voltage and current as well as inscribed within the characteristic curve, has greatest area.

5.3.7   FILL FACTOR

The fill factor indicates the quality of solar cell, that is, how much power or area of the characteristic curve is being used. In ideal case, the fill factor should be unity when the complete area between the characteristic curve and axes has been utilized that is, the product of V_{oc} and I_{sc}. The fill factor is defined as the ratio of peak power to the product of V_{oc} and I_{sc}:

$$FF = FF = \frac{Vm \times Im}{Voc \times Isc} \tag{5.5}$$

The typical value of fill factor is in the range of 0.5–0.83. The fill factor can be improved by: Increasing the photo current and decreasing the reverse saturation current of a solar cell; minimizing the internal series resistance; and maximizing the shunt resistance.

5.3.7.1   SOLAR EFFICIENCY

It is the ratio of maximum possible solar cell power output ($= V_m \times I_m$), which is converted to the solar energy supplied to the cell.

$$\text{Efficiency} = \frac{Vm \times Im}{\text{solar power}} \quad \text{or} = \frac{FF \times Voc \times Isc}{\text{solar power}} \tag{5.6}$$

Solar cell does not operate at the theoretical maximum efficiency because of several limitations. The efficiency of a solar cell varies from 12 to 15% only. Table 5.1 indicates the factors that limit the efficiency of a solar cell.

TABLE 5.1   Loss in Efficiency in a Solar Cell.

S. No.	Factors responsible for loss of efficiency	Percentage of loss
1	No photon absorption (photon energy less than forbidden energy)	23
2	Excess photon energy (photon energy more than forbidden energy)	33
3	Surface reflection	0.5
4	Voltage factor	18
5	Fill factor	5
6	Shading due to charge collection grid	0.05
7	Collection losses	5
8	Series resistance	0.5
	Total Loss	85.05
	Efficiency available in a solar cell = 100 − 85.05 =	14.95%

5.3.7.2   ENERGY LOSSES OF SOLAR CELL

The highest conversion efficiency of a solar cell is about 24%. Following factors lead to energy losses and limit the conversion efficiency of the cell:

5.3.7.2.1   Reflection Losses

Some of the incident radiation is lost due to reflection from the cell surface.

5.3.7.2.2   Incomplete Absorption

The energy of a photon is related to its wavelength (λ) by the equation: $E = \dfrac{h \times c}{\lambda}$, where h = Planck's constant (= 3×10^{-27} ergs); c = Velocity of light (3×10^{8} m/s). Using these values, we get: $E = \dfrac{1.24}{\lambda}$. The materials suitable for absorbing the energy of photons of sun light are: silicon, cadmium sulphide, and gallium arsenide. The difference between conduction and valence band is called band gap energy. Hence, photons having energy (E) larger than band gap energy (= 1.1 eV for silicon) will be absorbed in the cell material and will excite some of the electrons, thereby creating electron–hole pairs. Other photons of lower energy are wasted in generation of thermal energy. The higher is the band gap of the material, the greater is the wastage. The semiconductor with the energy gap of 0.9–1.1 eV would be best suited, and thickness required to absorb is about 300 μm.

5.3.7.2.3   Collection Losses

The collection efficiency is the ratio of the actual short circuit current density to the short circuit current density, which would be obtainable when no recombining takes place. The collection efficiency depends on: (a) the absorption characteristics of semiconductors which determine the generation of electron and hole pairs; (b) the junction depth; (c) the width of depletion layer; (d) the recombining rate of electron and holes; (e) the distance which carriers have to move for recombining; (f) the thickness of p and n regions; and (g) the existence and strength of any built-in electric field which help to accelerate carriers.

5.3.7.2.4 *Open Circuit Voltage*

The open circuit voltage is always less than the band gap energy due to lower level of illumination and doping of semiconductor, which lowers the potential difference at p–n junction. The increase in barrier potential increases V_{oc} but reduces I_{sc}. There is an optimum value of V_{oc} and I_{sc} for generation of maximum power output.

5.3.7.2.5 *Curve Factor*

The maximum power output is always less than the product of V_{oc} and I_{sc}. The characteristic curve does not have a rectangular shape. Hence the area of characteristic curves is always less than the product of V_{oc} and I_{sc}.

5.3.7.2.6 *Series Resistance Loss*

The voltage and current characteristic curve are flattened due to power loss resulting from series resistance. The output power decreases as the area under the characteristic curve reduces.

5.3.7.2.7 *Thickness of Cell*

Photons of high energy can pass through the cell material without any absorption if thickness is inadequate. A reflecting back ohmic contact is generally provided to enhance the absorption of high-energy photons.

5.3.7.3 *FACTORS AFFECTING THE OUTPUT OF A SOLAR CELL*

1. Sun light (irradiance): Solar cell output depends on the level of light or solar radiation flux falling on the solar cell surface.
2. Temperature of solar cell: Module temperature affects the output voltage inversely. Higher module temperature reduces the voltage by 0.04–0.1 V for every one degree Celsius rise in temperature. Air should be allowed to circulate behind the back of each module so that its temperature does not rise causing the increase of its output.

5.3.7.4   MAXIMIZING THE PERFORMANCE

The performance of a solar cell can be increased by taking the following steps:

a. Maximizing V_{oc} and I_{sc}: The efficiency of solar energy conversion depends upon V_{oc} and same time I_{sc} depends upon photocurrent and V_{oc} depends upon the ratio of I_{sc} to I_{o}.
b. Low series resistance: It will give high fill factor that is more output power possible as the area of characteristic curve increases. Reduction of resistance requires high doping of semiconductor.
c. High shunt resistance: Shunt resistance can be increased by preventing any leakage occurring at the perimeter of the cell. This is achieved by passivating the surface of the solar cell.
d. Optimum solar cell size: As the area of the solar cell increases, it becomes difficult to maintain the homogeneity of the material in solar cell and performance of the cell reduces.

5.3.7.5   NUMERICAL EXAMPLES

5.3.7.5.1   Numerical 1

Calculate the range of wavelength of solar radiation capable of creating electron–hole pair in silicon having energy gap of 1.12 eV,

$$E = hc/\lambda = \frac{1.24}{\lambda} \text{ or } \lambda = \frac{1.24}{1.12} = 1.11 \ \mu m.$$

5.3.7.5.2   Numerical 2

Considering the solar radiation of 800 W/m², find the area of PV cells needed to generate enough electric power to run (a) a desktop computer using 200 W (b) an electric geyser using 1 kW, and (c) a toaster using 300 W. Assume the efficiency of PV to be 25%.
    The PV cell power output = 0.25 × 800 = 200 W.

5.3.7.5.2.1   Case 1: Desktop Computer
Power of appliance required = 200 W; Area of PV cells = 200/200 = 1m².

5.3.7.5.2.2   Case 2: Electric Geyser

Power required = 1000 W; Area of PV cells = 1000/200 = 5m².

5.3.7.5.2.3   Case 3: Toaster

Power required = 300 W; Area of PV cells = 300/200 = 1.5m².

Total load = 200 + 1000 + 300 = 1500 W and total area of PV cells = 1 + 5 + 1.5 = 7.5m².

5.3.8   DESIGN OF A SPV SYSTEM

In order to design PV system, number of parameters about the component used in the system should be known. Following assumptions are made for design of a SPV system:

Inverter converts DC into AC power with an efficiency of about 90%. Battery charging and discharging cycle efficiency is about 90%. Also, not all the charge of a battery can be used. And one has to consider maximum depth of discharge of a battery. This can vary widely. Here, we are assuming 80% depth of discharge, meaning only 80% of the total capacity of the battery is useful.

The combined efficiency of inverter and battery will be calculated as:
= inverter efficiency × battery efficiency = 0.8 × 0.9 = 0.72 = 72%.

Battery voltage used for operation = 12 V; Battery capacity = 120 Ah.

Sunlight available in a day = 8 h/day (equivalent of peak radiation). *Note:* For most places in India, the sunlight available per day varies from 5 to 7 h/day equivalent of peak hours.

Operation of lights and fan = 6 h/day on PV panels.

PV panel power rating = 40 w_p.

In the operating condition, the actual output power of a PV module is less. Thus, a factor called "operating factor" is used to estimate the actual output from a PV module. The operating factor can vary between 0.60 and 0.90 (implying that output power is 60–80% lower than rated output power) under normal operating conditions, depending on temperature, dust on module etc. Thus, the actual output power of a 40 w_p PV panel = 0.75 (*operating factor*) × 40 = 30 W. Here, w_p implies peak power of a PV panel.

5.3.8.1   EXAMPLE: SPV WATER PUMPING SYSTEM

In general, a solar water pumping system consists of PV modules, motor, pump, and storage tank. The Figure 5.2 a shows the block diagram of SPV water pumping system. The storage tank can be thought of as an energy storage media like batteries. Therefore, the use of battery is not required for water pumping application. Also, a DC motor can directly be coupled with a SPV panel, avoiding the use of any inverter. An AC motor can be used with an inverter, which converts DC power of a PV model into AC power. Additionally a PV water pumping system can also have maximum power point tracking (MPPT) device to match the PV module output impedance with that of motor to extract maximum power throughout the day. Similar to solar home lighting system, a PV pumping system can be designed for sizes ranging from a very small water-pumping requirement for drinking water to large water volume requirement for irrigation purpose.

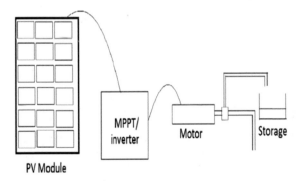

FIGURE 5.2   SPV water pumping system.

5.3.8.2   DESIGN OF SPV PUMPING SYSTEM

Design of SPV pumping system requires knowledge about how much water needs to be pumped, at what depth water should be pumped, how many solar panel will be required for a given water requirement, what should be the ratings of the motor used with PV panels etc. The overall design of the system can be done in the following steps:

Step 1: Determine the amount of water required per day.
Step 2: Determine the total dynamic head (TDH) for water pumping.

Step 3: Determine the hydraulic energy required per day (watt-hour (Wh)/day).

Step 4: Determine the solar radiation available at given location (in terms of equivalent of peak sunshine radiation, 1000 W/m²) in hours for which SPV module is characterized. Typically, this number is 5–8 under Indian conditions varying from season to season and location to location.

Step 5: Determine the size of SPV array and motor, consider motor efficiency, and other losses.

Before considering a case study, let us look at some of the definitions related to PV pumping:

5.3.8.3   TOTAL WATER REQUIREMENT (LITERS/DAY OR M³/DAY)

The size and cost of the system depends on the amount of water required per day. Solar pumping systems are designed to provide a certain quantity of water per day, where the daily water quantity required is sum of all requirements during 24 h. For more reliable design, worst case of water requirement should be considered (for instance, some people may need more water during summer). If the amount of water used everyday varies, then weekly average or monthly average can be taken for calculations.

5.3.8.4   TDH (METERS)

The TDH signifies the effective pressure at which pump must operate (Fig. 5.3). It primarily consists of two parameters: total vertical lift and total frictional losses.

The total vertical lift is a sum of elevation, standing water level, and drawdown as shown in Figure 5.3. The elevation is the height difference between the ground and the height at which water is discharged. Standing water level is the height difference between the ground surface and the water level in the well, when the well is in fully charged condition. And, drawdown is the height by which standing water level drops due to pumping, as shown in Figure 5.3.

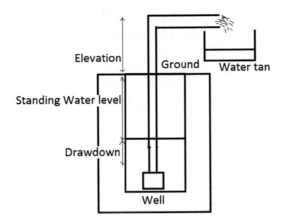

FIGURE 5.3   Pumping of water from a well.

5.3.8.5   FRICTIONAL LOSSES (EQUIVALENT METERS)

Frictional loss is the pressure required to overcome friction in the pipes from the water pump to the point of water discharge. It is given in equivalent meters and added to the total vertical lift for TDH calculation. The frictional loss depends on many factors like size of pipe, flow rate, type of fittings, number of bends etc. Usually nomographs are used to calculate the frictional loss. But if the water discharge point or tank is close to the well, then an approximation can be made. If the tank is within 10 m of the well, then frictional loss is taken as 5% of the total vertical lift.

5.3.8.5.1   Numerical Example 1

Design a PV system for pumping 25,000 l of water every day from a depth of about 10 m. Assume the following data:

- Amount of water to be pumped per day = 25,000 l = 25 m³.
- Total vertical lift = 12 m (5 m elevation, 5 m standing water level, 2 m drawdown).
- Water density = 1000 kg/m³.
- Acceleration due to gravity, g = 9.8 m/s².
- SPV module = 75 w$_p$.
- Operating factor = 0.75 (PV panel mostly does not operate at peak-rated power).

- Pump efficiency = 30% or 0.30 (can be taken between 0.25 and 0.4).
- Mismatch factor = 0.85 (PV panel does not operate at maximum power point).

Step 1: Determination total daily water requirement

Daily water requirement = 25 m³/day.

Step 2: Determination total dynamic head

Total vertical lift = 12 m.

Frictional losses = 5% of the total vertical lift = 12 × 0.05 = 0.6 m.

TDH = 12 + 0.6 = 12.6 m.

Step 3: Determine the hydraulic energy required per day

Hydraulic energy required to raise water level = Mass × g × TDH = density × volume × g × TDH = (1000 kg/m³) × (25 m³/day) × (9.8 m/s²) × 12.6 m × 1/3600

= 857.5 Wh/day.

Potential energy of the water is due to the elevation difference, which must be supplied by the pump.

Step 4: Determine solar radiation data

Solar radiation data in terms of equivalent peak sunshine radiation (1000 W/m²) varies between about 4 and 8 h. For exact hours, meteorological data should be used or can be estimated. Here, we use 6 h/day (peak of 1000 W/m² equivalent), actual day length is longer (this is equivalent of solar radiation of 180,000 Wh/month at a given location).

Step 5: Determine the number of PV panels and pump size

Total wattage of PV panel = (Total hydraulic energy)/(No. of hours of peak sunshine/day) = 857.5/6 = 142.9 W.

Considering system losses: Total PV panel wattage/Pump efficiency × Mismatch factor = 142.9/(0.3 × 0.85) = 560 W.

Considering operating factor for PV panel: Total PV panel wattage after losses/operating factor = 560/0.75 = 747.3 W.

Number of 75 w_p SPV panels required = 747.3/75 = 9.96 = 10 (approx.).

Power rating of the motor = 747.3/746 = 1 HP water.

In this way, a SPV water pumping system can be designed. This design example assumes the use of a DC motor. A system can also be designed for an AC motor but one must consider an inverter and its efficiency. Also, the cost of the SPV irrigation system can be estimated by considering cost for the individual components; cost of solar panel, cost of motor, and cost of pump and wiring cost.

5.3.8.5.2 *Numerical Example 2*

Calculate the number of PV module (each module of 40 W output) required for supplying power to operate a pump at 60% efficiency, if 60 m³ of water is to be lifted at a height of 5 m in an irrigation duration of 4 h.

The potential energy = PE = mgh; and W = ρvgh.

$$\text{Power} = \frac{\rho vgh}{t} = \frac{\left(1000 \text{ kg/m}^3\right)\left(60 \text{ m}^3\right)\left((9.8)\text{m}/\text{s}^2\right)(5 \text{ m})}{4\text{ h}\times 3600 \text{ s/h}} \quad 204.16\text{W}$$

Pump efficiency = 60%

$$\text{Power} = \frac{204.16}{0.6} = 340 \text{ W} \cong 360 \text{ W}$$

$$\therefore \frac{360}{40} = 9 \text{ Modules}$$

5.4 MATERIALS AND METHODS

The SPV micro-sprinkler irrigation system was designed and developed for cultivating tomato in one acre (0.4 ha) of land to achieve secured irrigation and to improve WUE mostly in vegetable cultivation. The details of the design and developments are discussed in this section.

During the year 2013, the experiments were carried out at the Central Farm of Orissa University of Agriculture and Technology (OUAT), Bhubaneswar, Odisha that lies at the latitude of 20° 15′ N and longitude of 85° 52′ E. The site falls under warm and humid climatic conditions. Tomato was cultivated both in *rabi* and summer seasons.

5.4.1   DESIGN OF SOLAR PHOTOVOLTAIC-POWERED MICRO-SPRINKLER IRRIGATION SYSTEM

5.4.1.1   WATER REQUIREMENT FOR VEGETABLE CROP

$$W_r = (A \times PE \times P_c \times K_c \times w_a)/E_u \qquad (5.1)$$

Where, W_r = peak water requirement (m³/day); A = crop area (m²); PE = pan evaporation rate (mm/day) converted to m/day; P_c = pan coefficient (0.7–0.9); K_c = crop coefficient (0.8–1); w_a = wetted area (%, 90% for micro-sprinkler irrigation); and E_u = emission uniformity of micro-sprinkler irrigation (approx. 0.8).

Using A = 4000 m²; PE = 8 mm/day; P_c = 0.85; K_c = 0.9; w_a = 0.9 and E_u = 0.8, we get: W_r = 27.54 m³/day (or 27,540 l/day). Taking irrigation interval to be 2 days, W_r = 27,540/2 = 13,770 l/day = 13.77 m³/day ≈ 14 m³/day.

5.4.2   SIZING OF PV MODULE FOR WATER REQUIREMENT OF 14 M³/DAY

$$E = (\rho\, g\, H\, V)/3.6 \times 10^6 \qquad (5.7)$$

Where, E = hydraulic energy required (kWh/day); ρ = density of water (1000 kg/m³); g = Gravitational acceleration (9.81 m/s²); H = total hydraulic head (m, 15 m in this case); V = volume of water required (14 m³/day in this case).

Using these values in Eq. (5.2), we get: E = 0.572 kWh/day = 572 Wh/day. Assuming actual sun shine hours in a day = 6 h, we have: Total wattage of PV module = 572/6 = 95.33 W. Following assumptions were made:

i.   Operating factor = 0.75–0.85 (PV panel mostly does not operate at peak-rated power).
ii. Pump efficiency = 70–80% (can be taken as 75%).
iii. Motor efficiency = 75–85% (can be taken as 80%).
iv. Mismatch factor = 0.75–0.85 (PV panel does not operate at maximum power).

Considering system losses, wattage requirement = (Total PV panel wattage)/(pump efficiency × mismatch factor) = (95.33)/(0.75 × 0.8) = 158.88 W.

Considering operating factor for PV panel = (Total PV panel wattage after losses)/(operating factor × motor efficiency) = (158.88)/(0.8 × 0.8) = 248.25 W.

Number of 75 w_p SPV panels required = 248.25/75 = 3.31 ≈ 4 modules

Power rating of motor = 248.25/746 = 0.33 hp

For micro-sprinkler irrigation system, the minimum rating of pump should be 1.5 hp. Hence, accordingly, the size of PV system needs to be decided.

5.4.2.1   SIZING OF PV SYSTEM FOR 1.5 HP RATING MOTOR

5.4.2.1.1   Battery Sizing

1.5 hp = 1119 W.

Daily water requirement for 1 acre = 14 m^3/day.

Hours of operation of motor per day for discharge of 14 m^3 water = 1 h.

Daily energy that needs to be supplied by battery is 1119 × 1 = 1119 Wh.

System voltage should be 24 V.

In SPV system, depth of discharge of battery may be from 70 to 80% (assume as 80%).

Hence, required charge capacity of batteries = 1119/24 = 46.6 Ah.

Total Ah capacity of battery = (Energy input to motor × No. of days of autonomy)/(depth of discharge × system voltage) = (1119 × 3)/(0.8 × 24) = 174.89 Ah.

Total number of batteries = (Total Ah capacity required)/(Ah capacity of one battery) = 174.89/100 = 1.74 ≈ 2 batteries.

For two batteries, 80 Ah + 80 Ah = 160 Ah is available instead of 140 Ah (1119 × 3/24).

These two batteries need to be connected parallel to get 160 Ah.

We have battery of 100 Ah with 12 V. Hence to get 24 V system voltages, two batteries should be connected in series. Hence, in total four batteries of 100 Ah with 12 V are required, two of them connected in series and two such series connected batteries can be connected in parallel.

5.4.2.2   PV SIZING TO MEET THE REQUIRED DAILY ENERGY REQUIREMENT

Normally battery efficiency varies from 80 to 90% (85% may be taken). Charge controller efficiency may be taken 90%.

The energy to be supplied by the PV system to the input of battery terminal = $(1119)/(0.85 \times 0.9) = 1462$ Wh.

Taking system voltage to be 24 V, Ah requirement = $1462/24 = 60.94$ Ah.

Taking daily sun shine hours to be 6 h, current requirement = $60.94/6 = 10.15$ A.

Normally a 75 w_p module generates 5 A current, hence number of modules = $10.15/5 = 2.03 \approx 2$ modules.

To get 10 A current and 24 V system voltage, we require 4 modules of 75 w_p. Two of them connected in series and two such series connected batteries to be connected in parallel.

5.4.2.2.1   Components Required for PV System

i.   Four batteries of 100 Ah 12 V.
ii. Four modules of 75 w_p.
iii. One charge controller.

5.4.3   COST OF SPV POWERED MICRO-SPRINKLER IRRIGATION SYSTEM

i.   Solar PV module (4×75 w_p) = 300 W at the rate of   = Rs. 18,000
     Rs. 60 per w_p
ii. Batteries (4×100 Ah, 12 V) at the rate of Rs. 5000 per   = Rs. 20,000
     battery
iii. Charge controller   = Rs. 3000
iv. 1.5 hp DC motor with pump set   = Rs. 12,000
v.   Sprinkler set up for 1 acre land   = Rs. 22,000
vi. Trolley rickshaw   = Rs. 13,000
vii. Pipes, fittings, wiring etc.   = Rs. 2000
     Total   = Rs. 90,000

The experimental set up was similar to the one indicated in Figure 5.4.

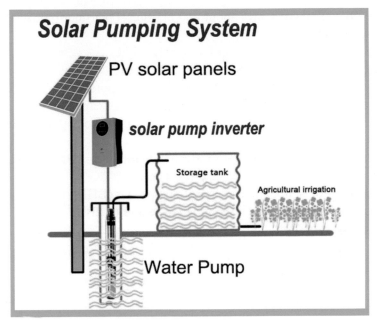

FIGURE 5.4   Solar photovoltaic integrated irrigation system.

5.4.3.1   INFORMATION FOR COST ANALYSIS

a.   Cost of 1.5 hp electric pump set = Rs. 10,000.
b.   Cost of 1.5 hp diesel pump set = Rs. 13,000.
c.   Cost of 1.5 hp PV powered pump set = Rs. 90,000.
d.   Prevailing interest rate = 10%.
e.   Efficiency of motor varies = 70–80% (70% here).
f.   Efficiency of pump varies = 70–80% (70% here).
g.   Efficiency of diesel engine = 30–40% (40% here).
h.   Useful life of PV panel = 20–25 years (22 years here).
i.   Useful life of diesel engine pump set = 8 years.
j.   Useful life of electric pump set = 8 years.
k.   Maintenance cost of PV system with sprinkler as 0.5% of total capital cost per year.
l.   Maintenance cost of diesel engine pump set as 10% of total capital cost per year.
m.   Maintenance cost of electric pump set as 10% of total capital cost per year.
n.   Annual working hours of diesel, electric pump sets and PV system = 500 h.

o. One hp engine consumes about 250 ml. diesel per hour (present cost of diesel Rs. 60/l).

p. One unit of electric energy (1 kWh) = Rs. 5.00.

q. Salvage value of diesel pump set = 20% of capital cost.

r. Salvage value of electric pump set = 20% of capital cost.

s. Salvage value of PV powered pump set = 5% of capital cost.

t. Operator's time spent in the proposed system = 1 h/day (labor charge Rs. 250/day).

u. Energy consumption (kWh) of electric pump set = (BHP)/(motor efficiency × pump efficiency) × 0.746 × 1 h.

v. Cost per hour of operation of diesel pump set = (BHP)/(motor efficiency × pump efficiency) × fuel consumed in liters/hour/BHP × cost of fuel/lit.

5.4.3.1.1 Hourly Operating Cost of PV Powered Water Pumping Device with Micro-Sprinkler System

5.4.3.1.1.1 Fixed Cost

i. Depreciation

D = (C − S)/(L × H), where C = capital cost; S = Salvage value; L = Useful life of device; H = Annual working hours (5.8)

Using the values we get, D = Rs. 7.77/h.

ii. Interest (I) = [(C + S)/(2)] × (Interest rate/100) × (1/H) = Rs. 9.45/h.

iii. Insurance and taxes and housing are not applicable.

Total fixed cost = 7.77 + 9.45 = Rs. 17.22/h.

5.4.3.1.1.2 Variable Cost

i. Fuel cost = Nil.

ii. Lubricants = Nil.

iii. Repair and maintenance = [(C) × (0.5/100)] × (1/H) = Rs. 0.9/h.

iv. Operator's wages Rs. 250/8 = Rs. 31.25/h.

Total variable cost = 0.9 + 31.25 = Rs. 32.15/h.

Total operation cost per hour = Total fixed cost/hour + Total variable cost/hour = 17.22 + 32.15 = Rs. 49.00/h.

5.4.3.1.2 Hourly Operating Cost of Diesel Pump Set

5.4.3.1.2.1 Fixed Cost

i. Depreciation,

$D = (C - S)/ (L \times H)$, where C = capital cost; S = Salvage Value; L = Useful life of device; H = Annual working hour.

Therefore: D = Rs. 2.6/h.

ii. Interest $(I) = (C + S)/(2) \times$ (Interest rate/100) $\times (1/H)$ = Rs. 1.56/h.
iii. Insurance and taxes and housing are not applicable.

Total fixed cost = 2.6 + 1.56 = Rs. 4.16/h.

5.4.3.1.2.2 *Variable Cost*

i. Fuel cost = $(1.5)/(0.4 \times 0.7) \times 0.25 \times 60$ = Rs. 80/h.
ii. Lubricants = 20% of cost of fuel = Rs. 16/h.
iii. Repair and maintenance = $(C) \times (10/100) \times (1/H)$ = Rs. 2.60/h.
iv. Operator's wages Rs. 250/8 = Rs. 31.25/h.

Total variable cost = 80 + 16 + 2.60 + 31.25 = Rs. 129.85/h.

Total operation cost per hour = Total fixed cost/h + Total variable cost/h = Rs. 134/h.

5.4.3.1.3 *Hourly Operating Cost of Electric Pump Set*

5.4.3.1.3.1 *Fixed Cost*

i. Depreciation

$D = (C - S)/ (L \times H)$ where C = capital cost; S = Salvage Value; L = Useful life of device; H = Annual working hour, Using the values of all necessary data, D = Rs. 2/h.

ii. Interest $(I) = (C + S)/(2) \times$ (Interest rate/100) $\times (1/H)$ = Rs. 1.2/h.
iii. Insurance and taxes and housing are not considered.

Total fixed cost = 2.0 + 1.2 = Rs. 3.2/h.

5.4.3.1.3.2 *Variable Cost*

i. Energy consumption (kWh) = $[(1.5)/(0.7 \times 0.7)] \times 0.746$ = 2.28 kWh.
ii. Electric energy cost = 2.28×5 = Rs. 11.40/h.
iii. Lubricants = 20% of cost of fuel = Rs. 2.28/h.
iv. Repair and maintenance = $(C) \times (10/100) \times (1/H)$ = Rs. 2/h.
v. Operator's wages Rs. 250/8 = Rs. 31.25/h.

Total variable cost = 11.40 + 2.28 + 2 + 31.25 = Rs. 46.93/h

Total operation cost per hour = Total fixed cost/hour + Total variable cost/hour = 3.2+ 46.93 = Rs. 50/h.

5.5   RESULTS AND DISCUSSION

Tomato is one of the most important and remunerative crops in Odisha and is grown in an area of 97,018 ha [2] covering 11.02% area of the total tomato cultivation in all India. It ranks second in the state in vegetable production. Odisha also ranks fourth among the tomato-producing states in India. It is considered as one of the most important supplementary sources of minerals and vitamins in human diet. However, targeted production and productivity is not achieved so far at par with the national level due to lack of assured irrigation facilitates both in _rabi_ and summer season.

The most prevailing variety of tomato that is, Utkal Kumari (BT-10) was selected for this study in order to evaluate the effectiveness of the developed SPV sprinkler irrigation device with respect to production and productivity, without depending upon conventional source of energy and flooded system of watering practice. The cost of cultivating tomato on one acre of land was calculated in order to know the annual profits out of it and its expected pay-back period. Similarly, the mitigation of greenhouse gases with the use of the developed set up was estimated and compared with traditional diesel and electric pump sets for its contribution in combating global warming and climate change and thus achieving sustainable agriculture.

5.5.1   COST–BENEFIT RATIO FOR TOMATO CULTIVATION ON 1 ACRE (0.4 HA) OF LAND

5.5.1.1   COST OF CULTIVATION OF TOMATO ON 1.0 ACRE LAND

Name of operation	Implements used	No. of operation	Labor h/Ac	Operation cost (Rs.)	Input (kg)	Cost of input (Rs.)	Total cost (Rs.)
Tillage	Tractor drawn rotavator	1 1	2 1	1200 600	–	–	1800
Planking	Wooden planker (manual)	1	2	31.25/h	–	–	62.50
Seed (Hybrid)							500
Planting (manual)		1	16	31.25/h			500
Fertilizer	FYM Gromer Potash	Once Twice Twice			1 Tractor load 100 kg 100 kg	4000 2500 2500	4000 2500 2500

(Continued)

Name of operation	Imple- ments used	No. of opera- tion	Labor h/Ac	Opera- tion cost (Rs.)	Input (kg)	Cost of input (Rs.)	Total cost (Rs.)
Interculture	Manual	Thrice	40	31.25/h			3750
Plant protection	Knapsack sprayer	Thrice	2	31.25/h	Pesticides	4000	4187
Irrigation	Solar PV powered sprinkler system	45 (2 days interval)	1	49/h			2205
Harvesting	manual	Twice/ week	120/ month				3750
Miscellaneous							4000
			Total cost of cultivation, Rs./acre = 29,754 ≈				30,000

5.5.1.2   BENEFIT

Without assured irrigation, production of tomatoes = 4000 kg/acre at the rate of Rs. 20/kg = Rs. 80,000.

With assured irrigation, production of tomatoes with 15% increase in yield = 4600 kg/acre at the rate of Rs. 25/kg = Rs. 115,000.

Net gain = Rs. 115,000 – Rs. 30,000 = Rs. 85,000 (in *rabi* season).

Net gain = Rs. 115,000 – Rs. 30,000 = Rs. 85,000 (in summer season).

Considering tomato cultivation in both the seasons in a year with assured irrigation, total gain = Rs. 170,000/annum.

Monthly income from tomato cultivation with assured irrigation = Rs. 14,167 ≈ Rs. 15,000 per month.

Simple payback period = (Initial investment cost)/(Net annual gain) = 90,000/1,70,000 = 0.5 years or 6 months.

5.5.2   MITIGATION OF CO_2 EMISSION BY USE OF SOLAR PHOTOVOLTAIC POWERED WATER PUMPING SYSTEM (SPVWPS)

Diesel and electricity are the two mostly used fuels to operate diesel and electric pump sets for water pumping in irrigating cultivable lands in state of Odisha. Burning of diesel in the internal combustion engines and generation of electricity in power plants contribute a lot in the emission of greenhouse gases to the atmosphere causing more to the present concern of global warming and climate change. This may be due to the strong initiatives being taken by the GOI to achieve more areas under assured irrigation. The replacement of diesel and electric pump sets with a reliable SPVWPS particularly in the irrigation sector can reduce the emission of greenhouse gases to the atmosphere. The existing diesel and electric pump sets in Odisha is 2.47×10^5 and 1.38×10^5, respectively, in the power rating range of 1–5 hp. Taking the average power rating of both diesel and electric pump sets as 3 hp, the amount of emission of CO_2 is as follows;

a. One hp engine consumes about 250 ml of diesel per hour.
b. Burning of 1 l of diesel releases 3 kg of CO_2 to the atmosphere [10].
c. The average carbon dioxide emission for electricity generation from coal based thermal power plant is approximately 1.58 kg of CO_2/kWh at the source.
d. Annual working hours of diesel and electric pump sets are assumed as 500 h.
e. Annual CO_2 emissions from 2.47×10^5 diesel pump are 300 million kg in Odisha.
f. Annual CO_2 emissions from 1.38×10^5 electric pump are 250 million kg in Odisha.
g. Total annual CO_2 emission can be mitigated by 550 million kg with the replacement of existing diesel and electric pump sets in Odisha by the adoption of SPV-powered system in the irrigation sector.
h. Total annual electrical energy consumption from 1.38×10^5 electric pump sets can be saved in the tune of 15×10^7 kWh (saving around 150 million units of electricity costing about Rs. 750 million/annum).
i. Total annual diesel consumption from 2.47×10^5 diesel pump sets can be saved in the tune of 10×10^7 l of diesel (saving around Rs. 6000 million/annum).

5.6 CONCLUSIONS

Sustainable energy source along with adoption of possible water management practices may be achieved with the help of SPV micro-irrigation system in order to solve the problem of inadequate availability of two critical inputs, such as energy and water for assured irrigation in agricultural sector. Micro-irrigation method through sprinkler system may also be an added advantage if integrated with the SPV device to achieve judicious utilization of water. Hence, use of SPV system may be a sustainable proposition of energy source for water pumping to achieve assured irrigation in the state. The findings of this study can give an insight to the farming community of the state to go for adopting the technology to strengthen their agricultural production system with secured availability of energy and water. The conclusions of this study are as follows;

a. Wide popularization of SPV-powered water pumping system for achieving assured irrigation through sustainable energy source.
b. Monthly income of Rs. 15,000/- throughout the year is possible by adopting remunerative tomato cultivation on 1 acre of land during *rabi* and summer seasons only.
c. The small and marginal farmers of the state may be attracted to adopt SPV-powered water pumping system as the hourly operating cost is lowest that is, Rs. 49/h followed by Rs. 50/h for electric pump set and Rs. 134/h for diesel pump set.
d. The existing area under vegetable cultivation in the state may be enhanced by adopting vegetable cultivation in the unutilized land mostly during summer season due to the assured irrigation facility through SPV system.
e. The developed set up may also be utilized for irrigating land in rainy season in case of irregular rainfall.
f. Pay-back period of the developed set up is only ½ year, due to which it may be easily accepted by the small and marginal farmers of the state in spite of its high initial cost.
g. Total annual CO_2 emissions can be mitigated by 550 million kg with the replacement of existing diesel and electric pump sets in Odisha by the adoption of a reliable SPV-powered system in irrigation sector.
h. Total annual electrical energy consumption from 1.38×10^5 electric pump sets can be saved in the tune of 15×10^7 kWh (saving around 150 million units of electricity costing about Rs. 750 million/annum).

Total annual diesel consumption from 2.47×10^5 diesel pump sets can be saved in the tune of 10×10^7 l of diesel (saving around Rs. 6000 million/annum).

KEYWORDS

- **cost–benefit ratio**
- **cost analysis**
- **diesel set**
- **electric motor**
- **energy security**
- **fixed cost**
- **flood irrigation**
- **greenhouse gases mitigation**
- **India**
- **micro-irrigation**
- **micro-sprinkler**
- **operational cost**
- **pay-back period**
- **pollution**
- **solar photovoltaic system**
- **tomato cultivation**
- **water use efficiency**

REFERENCES

1. Andrada, P.; Castro, J. *Solar Photovoltaic Water Pumping System Using a New Linear Actuator*; Grup d'Accionaments Electrics Amb Commutació Electrònica (GAECE): England, 2008.
2. Anonymous. *Agricultural Statistics*; Directorate of Agriculture and Food Production: Govt. of Odisha, 2012.
3. Anonymous. *Economic Survey of Odisha*; Planning & Coordination Department: Govt. of Odisha, 2013.
4. Anonymous. *Solar Policy*; Science and Technology Department: Govt. of Odisha, 2013.
5. Gopal, C.; Mohanraj, M.; Chandramohan, P.; Chandrasekhar, P. Renewable Energy Source Water Pumping Systems—A Literature Review. *Renew. Sust. Energ. Rev.* **2013,** *25,* 351–370.
6. Goyal, M. R. *Research Advances in Sustainable Micro Irrigation*; Apple Academic Press Inc: Oakville, ON, Canada, 2015; Vol. 1–10.

7. Hamidat, A.; Benyoucef, B.; Hartani, T. Small-Scale Irrigation with Photovoltaic Water Pumping System in Sahara Regions. *Renew. Energ.* **2003,** *28,* 1081–1096.

8. Kelley, L. C.; Gilbertson, E.; Sheikh, A.; Steven, D. E.; Dubowsky, S. On the Feasibility of Solar-Powered Irrigation. *Renew. Sust. Energ. Rev.* **2010,** *14,* 2669–2682.

9. Khatib, T. Design of Photovoltaic Water Pumping System at Minimum Cost for Palestine: A Review. *J. Appl. Sci.* **2010,** *10*(22)*,* 2773–2784.

10. Manfredi, S.; Tonini, D.; Christensen, T. Land Filling of Waste: Accounting of Greenhouse Gases and Global Warming Contributors. *Waste Manage. Res.* **2009,** *27,* 825–836.

11. Meah, K.; Sadrul, U.; Steven, B. Solar Photovoltaic Water Pumping-Opportunities and Challenges. *Renew. Sust. Energ. Rev.* **2008,** *12,* 1162–1175.

12. Meah, K.; Steven, F.; Sadrul, U. Solar Photovoltaic Water Pumping for Remote Locations. *Renew. Sust. Energ. Rev.* **2008,** *12,* 472–487.

13. Narela, P. D; Rathore, N. S.; Kothari, S. Study of Solar PV Water Pumping System for Irrigation of Horticulture Crops. *IJESI.* **2013,** *2*(12)*,* 54–60.

14. Sako, K. M.; Guessan, Y.; Diango, A. K.; Sangare, K. M. Comparative Economic Analysis of Photovoltaic, Diesel Generator and Grid Extension in Cote D'ivoire. *Asian J. Appl. Sci.* **2011,** *4,* 787–793.

CHAPTER 6

THERMAL MODELING OF A SOLAR GREENHOUSE FOR WATER SAVING AND SUSTAINABLE FARMING

M. K. GHOSAL

Department of Farm Machinery and Power, College of Agricultural Engineering and Technology, Orissa University of Agriculture and Technology, Bhubaneswar 751003, Odisha, India

E-mail: mkghosal1@rdiffmail.com

CONTENTS

ABSTRACT

The crop yield depends upon the environmental parameters, such as: air temperature, relative humidity, carbon dioxide concentration, soil temperature and moisture content to the soil, etc. The natural environment and input availability may not be optimum for a given crop. The congenial environment for a crop can be artificially maintained by means of a plastic covering structure like greenhouse. Greenhouses are available in various shapes and sizes suitable for different climatic zones. A greenhouse is an expensive option for rural farmers in India. Selecting a greenhouse that will perform most efficiently depends on many factors. A well-designed greenhouse is able to maintain a required environment inside its enclosure for healthy growth of plants resulting in better yield.

In this chapter, a mathematical model was developed for a greenhouse considering the effects of evaporative, radiative and conductive losses from the plants and the floor to predict the performance of a particular greenhouse in terms of various design, and climatic parameters. Experimental validation of the developed model was also done for typical days in the months of November 2012–February 2013 with the cultivation of one off-season okra in a low-cost and naturally ventilated greenhouse with shading nets in warm and humid climatic condition. Predicted values of air and plant temperatures were in close agreement with experimental values.

6.1   INTRODUCTION

The process of breaking complex system down into simpler elements and subjecting them to scrutiny is called analysis. In engineering, systems are broken down into subsystems, and subsystems into components. Everything is broken down until it becomes simple enough to be understood and to be converted into mathematical expressions that describe its properties, behavior, or function. The process of describing something physical by mathematical formulae is called analytical or mathematical modeling [3]. Hence, modeling is the representation of a system, process, or phenomenon occurring in real world situation and is expressed in mathematical form. The advantage of a model is two-fold: (a) by studying the model, it is possible to figure out how to change, improve, and optimize the design and (b) the performance and behavior of a design under a great variety of operational and environmental conditions can be predicted analytically long before the design is actually constructed and taken into service.

Thermal modeling of a controlled environment greenhouse is required to optimize the various parameters involved in either heating or cooling of greenhouse. The modeling can also be used to optimize greenhouse air temperature (one of the important constituents of the environment inside greenhouse) for enhancing production of a crop from greenhouse for a given thermal capacity. Thermal modeling requires basic energy balance equation for different components of greenhouse system for a given climatic (solar radiation, ambient air temperature, relative humidity, wind velocity, etc.), and design (volume, shape, height, orientation, latitude, etc.) parameters [5, 6, 10, 11, 13, 17].

To facilitate the modeling procedure, a greenhouse is considered to be composed of a number of separate but interactive components: the greenhouse cover, the floor, the growing medium, enclosed air, and the plant. The crop productivity depends on the proper environment and more specifically on the thermal performance of the system. The thermal performance of a greenhouse can be studied with the help of a mathematical model with suitable assumptions.

Energy balance equations have been derived to formulate the model, which permits the prediction of environmental conditions in a greenhouse from outside atmospheric conditions. Basic knowledge of heat and mass transfer is therefore of great importance in evaluating the thermal performance for heating and cooling operations of a greenhouse under given climatic conditions. The transfer of heat energy occurs as a result of temperature difference and mass transfer takes place in the form of evaporative heat transfer. Heat is transferred in four ways: thermal radiation, conduction, convection and evaporation, or transpiration. These modes of heat transfer occur frequently in nature and are governed by different laws.

The different processes of heat transfer occurring inside the greenhouse have been discussed in this chapter. The heat transfer in greenhouse system may also be classified as external and internal heat transfer. The external heat transfer processes that occur from roofs, walls, and ground to the outside greenhouse enclosure are through convection, radiation, and conduction. Heat transfer occurring within the greenhouse is referred to as internal heat transfer, which comprises of convection, radiation, and evaporation. In this chapter, an attempt has been made to develop a mathematical model based on energy balance equations for each component of the greenhouse. The mathematical model developed has then been validated by the recorded experimental findings.

6.2   SUSTAINABLE FARMING FOR SMALL AND MARGINAL FARMERS THROUGH LOW-COST GREENHOUSE

Growth of population and less availability of required food materials like vegetables (the important sources of vitamins, minerals, fiber, etc.) have become global concerns. In most of the developing countries, traditional open field cultivation is not able to maintain the sustainable vegetable production. Because open field agricultural practices only control the nature of the root medium through tillage operation, fertilizer application, and irrigation. They do not ensure the control on the environmental parameters, such as sunlight, air composition, and temperature that regulate the growth of plant. Hence, a large number of winter vegetables cannot be grown locally during summer period and have to be transported from the long distance places as per the needs of the consumers. The same practice also happens for summer crops during winter period. To meet the increased demand of off-season vegetables, greenhouse technology appears to be a promising alternative.

The demand of fresh as well as good quality vegetables at global level is also increasing. This calls for increasing productivity at a higher rate. The increased demand cannot be met through the traditional method of agricultural production. It necessitates improved and new alternative technologies to enhance production under normal as well as adverse climatic conditions and to bridge the gap between demand and existing production of vegetables.

Greenhouse, in this regard, helps to create favorable conditions where sustainable cultivation of vegetables is made possible throughout the year or part of the year as per the requirement. It not only creates suitable environment for the plants but also encourages proper growth and fruiting as compared to open field cultivation. The greenhouse technology has also tremendous scope especially for production of hybrid seeds, nursery raising [16], ornamental plants, medicinal plants, which fetch more prices in markets.

The control of various environmental parameters inside the greenhouse, suitable for favorable growth of plant can be studied mathematically by developing a suitable thermal model, which is required to optimize those parameters involved in either heating or cooling of greenhouse. The greenhouse environment is represented by a group of spatial average values of climatic factors, such as radiation, temperature, humidity, and carbon dioxide concentration, which effect the plant growth and development. Neither the plant architectural parameters nor the climatic variable is homogeneous in greenhouses.

The assemblage of environmental factors surrounding the living plants in a greenhouse is termed as greenhouse microclimate. It is this microclimate that directly governs energy and mass exchanges and influences plant metabolic activities. Greenhouse microclimate is influenced by both greenhouse macroclimate and the physical state (geometrical parameters, thermal conditions, etc.) of plant elements subjected to it [18]. The difference between greenhouse climate and outside environment is mainly caused by two mechanisms:

a. The first is the **enveloping of air,** which is stagnant in the greenhouse due to the enclosure. So the exchange of the greenhouse air with the surrounding (outside) air is strongly decreased compared to that of the air without envelope. Moreover, the local (greenhouse) air velocities are small compared to that in the open condition. The reduction of the air exchange (or ventilation) directly affects the energy and mass balances of the greenhouse air while the smaller local air velocities affect the exchange of energy, water vapors and carbon dioxide between the greenhouse air, and the greenhouse inventory (crop, soil surface, enclosure, and heating system).
b. The second is the **mechanism of radiation,** in which the inward short-wave radiations (direct from the sun and scattered from the sky and clouds) are intercepted by the opaque and transparent components of the greenhouse, while the long-wave radiative exchange between inside and outside environment is affected due to the radiative properties of the covering materials.

The greenhouse microclimate quantitatively describes the energy and mass transfer processes within the canopy, the exchange processes between air and plant elements and other surfaces, and the ways in which plants respond to environmental factors [18]. The greenhouse microclimate may be affected by the orientation of the greenhouse, latitude [12], area of the greenhouse, canopy area inside the greenhouse, bare soil surface area inside the greenhouse, structural design (shape and size) of the greenhouse, properties of the material used for the construction of the greenhouse, etc. The microclimate will also depend on the ventilation system provided inside the greenhouse that is, whether it is naturally ventilated or fans are provided with or without cooling pads [14]. Hence, various models need to be developed to predict the greenhouse microclimate under various climatic conditions. The controlled environment through solar greenhouse is therefore, the right alternative for the higher production and productivity of off-season vegetables by maintaining the required environmental conditions for their growth and yield. So there is a need to increase the temperature of air inside the enclosure for safe growing of off-season vegetable in winter season as well as decrease of temperature in the pre-summer and summer periods. As greenhouse allows faster temperature increase during sunny day and slower temperature decrease in night hours, a suitable structure for off-season cultivation of these vegetables needs to be thought of for use throughout the year. But higher operating cost of high-tech controlled greenhouse will be a constraint for popularization of this technique in the developing countries where more than 85% of farming community is of small and marginal category [1, 2]. Hence, there is a need to study the suitability of low-tech naturally ventilated greenhouse for off-season cultivation of vegetables, which would become more remunerative for sustainable farming especially in case of small and marginal farmers [15].

6.3   EXPERIMENTAL SET-UP

Two semi-circular shaped greenhouses (Figs. 6.1 and 6.2) located adjacent to each other in the nursery site of the Department of Horticulture, Orissa University of Agriculture and Technology, Bhubaneswar, Odisha, India were used for the study and developing the thermal model. The floor space of each of the experimental greenhouse was 4×12 m (48 sq m) and oriented in east–west direction. One greenhouse was covered with ultra violet (UV) low-density polyethylene (LDPE) film of 200 micron thickness. Similarly, the other greenhouse of same size with same UV film of 200 micron thickness was covered with a Netlon make shading net of 50% to study the performance of the greenhouse with shading net. The shade net used in the present experimental greenhouse was of green colored UV resistant high-density polyethylene (HDPE) material. The durability of the shade net is about 4–5 years. The cost of shade net was Rs. 18 sq m. The experimental observations were taken during the year 2012–2013.

The site is situated at $20^0 15`$ N latitude and $85^0 52`$ E longitude with an elevation of 25.9 m above the mean sea level and nearly 64 km west of the Bay of Bengal and coming under the warm and humid climatic condition. The ambient air temperatures usually vary from 25 to 37 °C in summer, 24 to 32 °C in rainy and 17 to 27 °C in winter seasons. To ventilate the greenhouse freely with the outside cool air for controlling the air temperature and humidity inside the greenhouse as per the requirement, the polythene sheet of both sides (north and south) of the greenhouse were rolled upward and downward to maintain the partial thermal environment. Constructional details of the greenhouse have also been shown in Figure 6.3. The crop grown in both the greenhouses was okra (Figs. 6.4 and 6.5).

FIGURE 6.1   Experimental greenhouse without shading net.

FIGURE 6.2   Experimental greenhouse with shading net.

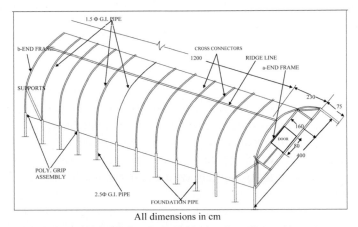

All dimensions in cm

FIGURE 6.3   Constructional details of semi-circular shaped greenhouse.

FIGURE 6.4   Okra plants under greenhouse with shading net.

FIGURE 6.5   Okra plants under greenhouse without shading net.

6.4   DEVELOPMENT OF THE MATHEMATICAL MODEL FOR GREENHOUSE UNDER STUDY

To develop a thermal model, the greenhouse is considered to be composed of a number of separate but interactive components: the greenhouse cover, the floor, the growing medium, enclosed air, and the plant. The crop productivity depends on the proper environment and more specifically on the thermal performance of the system. The thermal performance of a greenhouse can be studied with the help of a mathematical model through suitable assumptions. Energy balance equations have been derived to formulate the model, which permits the prediction of environmental conditions in a greenhouse from the outside atmospheric conditions. The energy balance equations for various components of greenhouse can be written on the basis of the following assumptions:

a. Analysis is based on quasi-steady state condition [19].
b. Storage capacity of greenhouse cover materials is neglected.
c. Absorptivity and heat capacity of air is neglected.
d. Heat flow in the ground is one dimensional.
e. Thermal properties of plants in the greenhouse are nearly same as those of water.
f. There is no radiative heat exchange between the walls and roofs of greenhouse due to negligible temperature differences.

The energy balance equations for various components like plant mass, floor, and air of greenhouse (Fig. 6.6) can be written as follows:

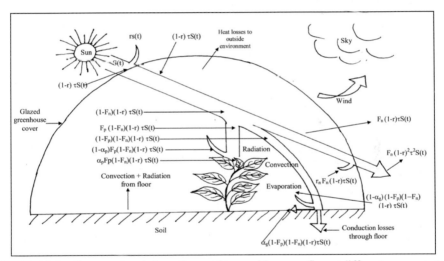

FIGURE 6.6 Cross sectional view of a greenhouse with energy flow at different components.

6.4.1 GREENHOUSE PLANT MASS

$$\alpha_p \, F_p \, (1 - F_n)(1 - r)\, \tau S(t) = M_p \, C_p \, \frac{dT_p}{dt} + h_p \, (T_p - T_r)\, A_p \qquad (6.1)$$

Rate of thermal energy absorbed by plant mass in the greenhouse	= Rate of thermal energy stored in the plant mass	+ Rate of thermal energy transferred from plants to greenhouse air.

where:

$$h_p = h_{cp} + h_{ep} \; ; \; h_{cp} = h_i \text{ and } h_{ep} = 0.016 \, h_{cp} \left[\frac{P(T_p) - \gamma \, p(T_r)}{T_p - T_r} \right]$$

6.4.2   GREENHOUSE FLOOR

$$\alpha_g (1 - F_p)(1 - F_n)(1 - r) \tau S(t) = h_{gr} \left(T \big|_{x=0} - T_r \right) A_g + \left(-K_g \frac{\partial T_g}{\partial x} \bigg|_{x=0} \right) A_g \quad (6.2)$$

Rate of thermal energy = Rate of thermal energy + Rate of thermal
transferred from the   transferred from ground   energy conducted into
greenhouse floor   surface to greenhouse   the greenhouse floor

Rate of thermal energy conducted into the ground is equal to the rate of overall heat transfer from floor to the higher depth of ground. Hence,

$$-K_g \frac{\partial T_g}{\partial x} \bigg|_{x=0} = h_{g\infty} \left(T \big|_{x=0} - T_\infty \right)$$

Also at higher depth, the temperature of ground is nearly equal to the ambient temperature. Therefore,

$$-K_g \frac{\partial T_g}{\partial x} \bigg|_{x=0} = h_{g\infty} \left(T \big|_{x=0} - T_\infty \right) = h_{g\infty} \left(T \big|_{x=0} - T_a \right)$$

Therefore, Eq 6.2 becomes

$$\alpha_g (1 - F_p)(1 - F_n)(1 - r) \tau S(t) = h_{gr} \left(T \big|_{x=0} - T_r \right) A_g + h_{g\infty} \left(T \big|_{x=0} - T_a \right) A_g \quad (6.3)$$

6.4.3   GREENHOUSE ENCLOSED AIR

$$(1 - \alpha_p) F_p (1 - F_n)(1 - r) \tau S(t) + (1 - \alpha_g)(1 - F_p)(1 - F_n)(1 - r) \tau S(t)$$
$$+ r_n F_n (1 - r) \tau S(t) + h_p (T_p - T_r) A_p + h_{gr} \left(T \big|_{x=0} - T_r \right) A_g =$$
$$\left(\sum A_i U_i \right) (T_r - T_a) + 0.33 \, NV (T_r - T_a) \quad (6.4)$$

| Rate of thermal energy retained in greenhouse air after absorption by plant mass | + | Rate of thermal energy retained in greenhouse air after absorption in floor | + | Rate of thermal energy retained in greenhouse air after reflected from glazed north wall |

| + | Rate of thermal energy transferred from plants to greenhouse air | + | Rate of thermal energy transferred from floor to greenhouse air | = | Rate of overall heat loss between greenhouse air and ambient air | + | Rate of thermal energy lost due to ventilation |

Eqs 6.1, 6.3, and 6.4 can be solved for determining the temperature of greenhouse air (T_r) and plants (T_p). Eliminating $T|_{x=0}$ from Eq 6.3 and after rearranging, we get:

$$h_{gr}\left(T|_{x=0} - T_r\right)A_g = F_1 I_{eff} - A_g U_g (T_r - T_a)$$ (6.5)

where:

$$I_{eff} = \alpha_g (1 - F_p)(1 - F_n)(1 - r)\,\tau S(t); F_1 = \frac{h_{gr}}{h_{gr} + h_{g\infty}}$$

$$\text{and } U_g = \frac{(h_{gr})(h_{g\infty})}{h_{gr} + h_{g\infty}}$$

Substituting Eq 6.5 in Eq 6.4 and simplifying, Eq 6.4 can be written for the expression of T_r as:

$$T_r = \frac{I_{effA} + F_1 I_{effF} + A_p h_p T_p + Z T_a}{A_p h_p + Z}$$ (6.6)

where:

$$I_{effA} = (1 - \alpha_p) F_p (1 - F_n)(1 - r)\tau S(t) + (1 - \alpha_g)(1 - F_p)(1 - F_n)$$
$$(1 - r)\tau S(t) + r_n F_n (1 - r)\tau S(t) \text{ from Eq 6.4; and}$$
$$Z = A_g U_g + \left(\sum A_i U_i\right) + 0.33 NV$$

Substituting the expression of T_r from Eq 6.6 in Eq 6.1 and after simplification, Eq 6.1 can be written in the following first order differential equation.

$$\frac{dT_p}{dt} + aT_p = F(t) \tag{6.7}$$

where:

$$a = \frac{HZ}{M_pC_p}; F(t) = \frac{I_{effP} + H(I_{effA} + F_1 I_{effF} + ZT_a)}{M_pC_p}$$

$$H = \frac{A_p h_p}{A_p h_p + Z}$$

$$I_{effP} = \alpha_p F_p (1-F_n)(1-r)\tau S(t)$$

where:

$$S(t) = \sum_{i=1}^{8} A_i I_i = A_1 I_1 + A_2 I_2 + A_3 I_3 + A_4 I_4 + A_5 I_5 + A_6 I_6 + A_7 I_7 + A_8 I_8$$

$$\sum_{i=1}^{8} A_i U_i = A_1 U_1 + A_2 U_2 + A_3 U_3 + A_4 U_4 + A_5 U_5 + A_6 U_6 + A_7 U_7 + A_8 U_8$$

$$U_1 = U_2 = U_3 = U_4 = U_5 = U_6 = U_7 = U_8 = U \text{ ; and}$$

$$U = \left[\frac{1}{h_i} + \frac{1}{h_0}\right]^{-1}; \text{ and}$$

$$h_i = 5.7 + 3.8\, v = 5.7,$$

if velocity of air inside greenhouse is taken zero and $h_0 = 5.7 + 3.8\, v$

$$h_{g\infty} = \left[\frac{L_g}{K_g}\right]^{-1} \qquad h_{gr} = h_i$$

and $P(T_p) = \exp\left[25.32 - \dfrac{5144}{T_p + 273}\right]$ and $P(T_r) = \exp\left[25.32 - \dfrac{5144}{T_r + 273}\right]$

Analytical solution of Eq 6.7 can be written as:

$$T_p = \frac{\overline{F(t)}}{a}(1 - e^{-at}) + T_{p0}e^{-at}$$

(6.8)

where: T_{p0} is the temperature of plant at $t = 0$; and $\overline{F(t)}$ is the average value of $F(t)$ for the time interval between 0 and t; and a is constant during the time.

Once the numerical value of T_p is determined, then greenhouse air temperature (T_r) can be determined from the Eq 6.6.

6.5   ENERGY BALANCE EQUATIONS OF GREENHOUSE WITH SHADING NET

The assumptions made for writing the energy balance equations for greenhouse with shading net are same as mentioned in Section 6.4 of this chapter.

6.5.1   ENERGY BALANCE EQUATIONS DURING DAY TIME (9 A.M.–6 P.M.)

6.5.1.1   PLANT MASS

The energy balance equation of plant mass inside greenhouse with shading net is same as Eq 6.1.

6.5.1.2   GREENHOUSE FLOOR

The energy balance equation of greenhouse floor for greenhouse with shading net is same as Eq 6.3.

6.5.1.3   GREENHOUSE AIR

The energy balance equation of air inside greenhouse with shading net is same as Eq 6.4. The procedure of determining T_r and T_p is same as mentioned above. The only differences are in the expressions for $S(t)$ and $(\Sigma A_i U_i)$, which are as follows:

$$S(t) = \sum_{i=1}^{8} A_i I_i \tau_i, \text{ or}$$

$S(t) = A_1 I_1 \tau_1 + A_2 I_2 \tau_2 + A_3 I_3 \tau_3 + A_4 I_4 \tau_4 + A_5 I_5 \tau_5 + A_6 I_6 \tau_6 + A_7 I_7 \tau_7 + A_8 I_8 \tau_8,$ and
$\tau_1 = \tau_2 = \tau_3 = \tau_4 = \tau_5 = \tau_6 = \tau_s$ (transmittivity of shading net), and
$\tau_7 = \tau_8 = \tau$ (transmittivity of greenhouse cover).

$$\sum_{i=1}^{8} A_i U_i = A_1 U_1 + A_2 U_2 + A_3 U_3 + A_4 U_4 + A_5 U_5 + A_6 U_6 + A_7 U_7 + A_8 U_8$$

$$U_1 = U_2 = U_3 = U_4 = U_5 = U_6 = U_s \text{ and } U_7 = U_8 = U \text{ and } U_s = \left[\frac{1}{h_i} + \frac{L_s}{K_s} + \frac{1}{h_0} \right]^{-1}$$

6.5.2   ENERGY BALANCE EQUATIONS DURING NIGHT TIME (6 P.M.–9 A.M.)

$S(t) = 0$ in Eqs 6.1, 6.3, and 6.4.

6.5.2.1   PLANT MASS

$$h_p (T_p - T_r) A_p = M_p C_p \frac{dT_p}{dt} \tag{6.9}$$

6.5.2.2   GREENHOUSE FLOOR

$$h_{gr} \left(T\big|_{x=0} - T_r \right) A_g = h_{g\infty} \left(T_a - T\big|_{x=0} \right) A_g \tag{6.10}$$

6.5.2.3 GREENHOUSE ENCLOSED AIR

$$h_{gr}\left(T\big|_{x=0} - T_r\right)A_g + h_p(T_p - T_r)A_p = \left(\sum A_i U_i\right)(T_r - T_a)$$
$$+ 0.33\,NV(T_r - T_a) \tag{6.11}$$

Eliminating $T\big|_{x=0}$ from Eq 6.10 and after rearrangement,

$$h_{gr}\left(T\big|_{x=0} - T_r\right)A_g = A_g U_g(T_a - T_r) \tag{6.12}$$

Substituting Eq 6.12 in Eq 6.11 and simplifying, Eq 6.11 can be written for the expression for T_r is as follows:

$$T_r = \frac{A_p h_p T_p + Z T_a}{A_p h_p + Z} \tag{6.13}$$

Substituting the expression of T_r from Eq 6.13 in Eq 6.9 and after simplification, Eq 6.9 can be written in the following first order differential equation:

$$\frac{dT_p}{dt} + aT_p = B(t) \tag{6.14}$$

where: $B(t) = \dfrac{HZ}{M_p C_p} T_a$

Analytical solution of Eq 6.14 can be written as

$$T_p = \frac{\overline{B(t)}}{a}(1 - e^{-at}) + T_{p0}e^{-at} \tag{6.15}$$

where: T_{p0} is the temperature of plant at $t = 0$, and $\overline{B(t)}$ is the average value of $B(t)$ for the time interval between 0 and t, and a is constant during the time.

Once the numerical value of T_p is determined, then greenhouse air temperature (T_r) during night time can be determined from the Eq 6.13. The temperatures of plant and air inside greenhouse during night time will be different from those in day time.

6.6   COMPUTATIONAL PROCEDURE AND INPUT PARAMETERS

The mathematical model was solved with the help of the computer program in MATLAB. Numerical calculations were made corresponding to the hourly variations of solar radiation and ambient air temperature for typical winter days of clear sunny days (November 5, 2012), (December 15, 2012), (January 12, 2013), and (February 4, 2013) of Bhubaneswar. Solar radiation falling on different walls and roofs of the greenhouse was calculated with the help of Liu and Jordan formula by using the beam and diffuse components of solar radiation incident on the horizontal surface [8]. The input parameters and design parameters used for experimental validation are given in Tables 6.1 and 6.2, respectively.

TABLE 6.1   Total Solar Radiation and Diffuse Radiation Available on Horizontal Surface, Solar Fraction on North Wall and Ambient Air Temperature for Typical Dates of Experiments.

Time	November 5, 2012				December 15, 2012			
	I_h	I_d	F_n	T_a	I_h	I_d	F_n	T_a
(Hour)	(W/m²)		–	(°C)	(W/m²)		–	(°C)
1	0	0	0	25.1	0	0	0	16.3
2	0	0	0	24.6	0	0	0	15.2
3	0	0	0	24.3	0	0	0	14.5
4	0	0	0	24.1	0	0	0	14.2
5	0	0	0	24.0	0	0	0	14.0
6	105	105	0	24.3	105	105	0	14.8
7	187	131	0	24.7	130	121	0	15.3
8	411	161	0.39	25.5	303	145	0.4	16.8
9	595	170	0.40	26.5	489	162	0.45	19.2
10	762	172	0.32	27.8	636	171	0.37	22.6
11	805	174	0.22	29.1	707	172	0.38	25.3
12	843	180	0.17	30.6	739	174	0.32	28.5
13	797	171	0.20	31.8	717	173	0.30	29.7
14	728	170	0.31	32.1	605	164	0.28	20.1
15	583	166	0.39	31.8	470	156	0.38	29.6
16	481	152	0.38	31.3	282	136	0.42	29.0
17	211	149	0	31.1	107	99	0	28.3

TABLE 6.1 *(Continued)*

Time	November 5, 2012				December 15, 2012			
	I_h	I_d	F_n	T_a	I_h	I_d	F_n	T_a
(Hour)	(W/m²)		–	(°C)	(W/m²)		–	(°C)
18	109	109	0	30.2	85	85	0	26.7
19	0	0	0	29.3	0	0	0	24.5
20	0	0	0	28.3	0	0	0	22.8
21	0	0			0	0	0	21.1
22	0	0	0	26.6	0	0	0	19.6
23	0	0	0	25.9	0	0	0	18.5
24	0	0	0	25.9	0	0	0	17.5

TABLE 6.2 Design Parameters Used for Computation.

Parameter	Value	Parameter	Value	Parameter	Value
A_1, A_6	7.2 m²	h_p	30.25 W/m² °C	V	85.32 m³
A_2, A_5	9,6 m²	h_{gr}	5.7 W/m² °C	r_n	0.1
A_3, A_4	21.6 m²	K_g	0.52 W/m °C	α_g	0.4
A_7, A_8	7.11 m²	L_g	1.00 m	α_p	0.5
A_g	48 m²	M_p	215 kg	τ	0.5
A_p	110 m²	N	1–10	τ_s	0.1
C_p	4190 J/kg °C	h_{cp}	5.7 W/m² °C	–	–
F_n	0.08–0.45	h_{ep}	24.55 W/m² °C	–	–
F_p	0.3	r	0.2	–	–
h_i	5.7 W/m² °C	U	3.5 W/m² °C	–	–
h_0	9.5 W/m² °C	v	0.5–1.5 m/s	–	–

Hourly variations of air and plant temperatures for greenhouse without shading net and with shading net both in day time and night time were recorded during experimentation. For analysis of thermal environment of greenhouse, quasi-steady state method has been used. In order to verify the accuracy of the model, developed, the predicted values of temperatures of air, and plants inside greenhouse were validated against the experimental results for typical sunny days. The closeness of predicted and experimental values has been verified with the help of coefficient of correlation (r) and root mean square of percent deviation (e).

6.6.1   STATISTICAL TOOLS

6.6.1.1   COEFFICIENT OF CORRELATION (C) AMONG THE PREDICTED AND EXPERIMENTAL VALUES [3]

$$C = \frac{N' \sum X_i Y_i - (\sum X_i)(\sum Y_i)}{\sqrt{N' \sum X_i^2 - (\sum X_i)^2} \sqrt{N' \sum Y_i^2 - (\sum Y_i)^2}} \tag{6.16}$$

where: N' is the number of observations; X_i and Y_i are predicted and experimental values, respectively.

6.6.1.2   ROOT MEAN SQUARE OF PERCENT DEVIATION (E)

$$e = \sqrt{\frac{\sum (e_i)^2}{N'}}, \text{ where} \tag{6.17}$$

$$e_i = \left[\frac{X_{predicted(i)} - X_{experimental(i)}}{X_{predicted(i)}} \right] \times 100, \text{ and } N' \text{ is the number of observations.}$$

6.6.1.3   RATE OF HEAT LOSS THROUGH VENTILATION FROM GREENHOUSE ENCLOSURE TO AMBIENT AIR

$$\dot{q} = m_a C_a (T_r - T_a)$$

$$= \frac{V \rho_a}{t} C_a (T_r - T_a) = \frac{V \times 1.2}{\dfrac{3600}{N}} \times 1000 (T_r - T_a) = 0.33 \, NV \ (T_r - T_a)$$

where: $\dot{q}$ is in watts, density of air ρ_a = 1.2 kg/m³, specific heat of air C_a = 1000 J/kg °C, t is in seconds and t = 3600/N.

6.6.1.4 SOLAR FRACTION (F_n)

The concept of solar fraction is very important in thermal analysis of a greenhouse system. It gives an idea regarding the distribution of solar radiation in walls and floor of greenhouse as compared to total incoming solar radiation. Solar fraction is defined as the solar energy contribution to the total load requirement for a system. For an east–west oriented greenhouse in northern hemisphere particularly in winter period, most of the solar radiation falls on the south wall and is transmitted to the atmosphere through the glazed (greenhouse) cover. Hence, the fraction of solar radiation falling on north wall to the total solar radiation coming into the greenhouse is important in thermal model of a greenhouse. Therefore, solar fraction is defined as follows:

$$F_n = \frac{\text{Solar radiation available on north wall inside the greenhouse for a given time}}{\text{Total solar radiation entering into the greenhouse for same time}}$$

The solar fraction depends on solar altitude angle, angle of incidence, shape as well as size of greenhouse. This value is less than one and varies during sunshine hours in a day [9].

6.7 EXPERIMENTAL VALIDATION AND FINDINGS FROM THE GREENHOUSE UNDER STUDY

The experimental validation of the model developed has been carried out for typical days during November, 2012 through February, 2013. The experimental and the predicted values of plant and air temperatures (T_p and T_r) of greenhouse without shading net along with ambient air temperatures (T_a) for typical days (clear and sunny days) that is, on November 5, 2012; December 15, 2012; January 12, 2013; and February 4, 2013 have been shown from Figures 6.7 to 6.10. Similarly, the experimental and the predicted values of plant and air temperatures (T_p and T_r) of greenhouse with shading net for the above mentioned days have been shown from Figures 6.11 to 6.14.

It was observed from the figures that the predicted values of plant and greenhouse air temperatures were fairly close to the experimental values. These have been verified with the help of statistical analysis of r and e between the experimental and predicted values for both the greenhouses under study. From the values indicated in the figures, it was found that the r among the predicted and experimental values of plant and enclosed air temperatures for both the greenhouses varied from 0.94 to 0.97 and e varied from 5.89 to 10.70.

Due to incident solar radiation, the enclosed air temperature (T_r) and the plant temperature (T_p) began to rise from 10 a.m. and attained a value, which was beyond the favorable temperatures for okra plants. At this stage, the sides of the greenhouse were kept opened till 4 p.m. in order to facilitate the removal of excess heat energy from the greenhouse enclosure due to natural ventilation [4, 7]. For the natural convection mode, number of air changes per minute was experimentally obtained one, which has been taken into account in the model.

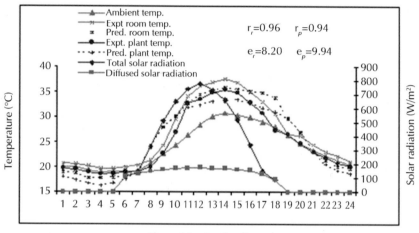

FIGURE 6.7 Hourly variations of plant and greenhouse air temperatures (experimental and predicted), ambient air temperature, and solar intensity of greenhouse without shading net on November 5, 2012 from 1 a.m. to 12 midnight.

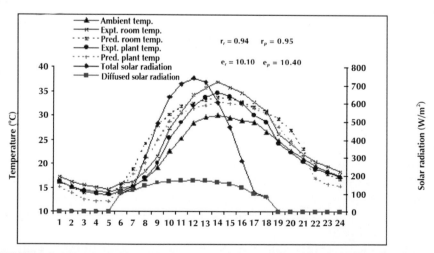

FIGURE 6.8 Hourly variations of plant and greenhouse air temperatures (experimental and predicted), ambient air temperature, and solar intensity of greenhouse without shading net on December 15, 2012 from 1 a.m. to 12 midnight.

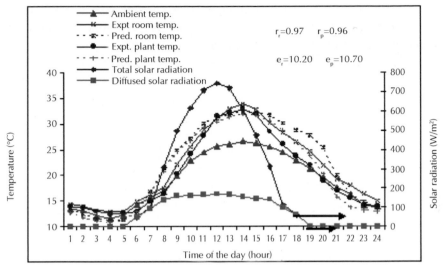

FIGURE 6.9 Hourly variations of plant and greenhouse air temperatures (experimental and predicted), ambient air temperature and solar intensity of greenhouse without shading net on January 12, 2013 from 1 a.m. to 12 midnight.

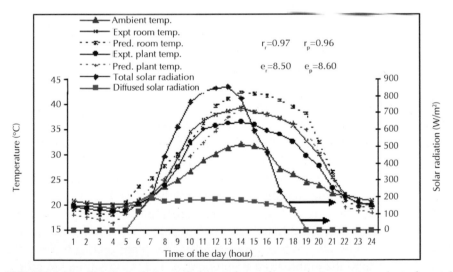

FIGURE 6.10 Hourly variations of plant and greenhouse air temperatures (experimental and predicted), ambient air temperature and solar intensity of greenhouse without shading net on February 4, 2013 from 1 a.m. to 12 midnight.

During peak sunny hours, the air temperature inside the greenhouse without shading net (Fig. 6.7–6.10) were 5–7 °C higher than the ambient air temperatures and during mid night hours the greenhouse air temperature was only 1–2 °C higher than the ambient air temperatures. The observed plant temperatures were 3–5 °C higher than the ambient air temperatures and 1–2 °C less than the greenhouse air temperature during peak sunny hours. During night hours, the plant temperatures were found to be 2 °C less than the ambient air temperature and 1–2 °C less than the greenhouse air temperatures. Natural ventilation was done from 10 a.m. to 4 p.m. to keep the greenhouse air temperature within 5–7 °C higher than the ambient air temperature to make it suitable for the crop growth inside the greenhouse. In the night hours, greenhouse cover maintained inside temperature 1–2 °C higher than the ambient air temperature during winter days. The ambient temperatures of air generally vary from 17 to 27 °C during winter season. Hence, it was difficult to maintain the temperatures ranges of 22–28 °C inside the greenhouse without shading net for favorable growth of the plants mainly okra.

Similarly, in case of greenhouse with shading net (Fig. 6.11–6.14), the enclosed air temperatures (T_r) and plant temperatures (T_p) attained the highest value at around 1–2 p.m., which were also not favorable for the growth of okra plants. At this stage the sides were kept opened for natural ventilation. The combined effect of natural ventilation and shading net reduced the temperatures of air inside the greenhouse to the extent favorable for the plants both in day time and during night time. During peak sunny hours, the greenhouse air temperature inside the shading net is only 1–2 °C higher than the ambient air temperature and 3–4 °C lower than the air temperature of greenhouse without shading net but in the night hours the inside temperature is 3–5 °C higher than the ambient air temperature and 2–3 °C higher than the air temperatures of greenhouse without shading net, which were more congenial for the growth of plants like okra (22–28 °C) during the winter days in the coastal region of Odisha. The ambient air temperatures during winter season usually vary from 17 to 27 °C. The observed plant temperatures inside the shading net during peak sunny hours is 1–2 °C lower than ambient air temperature, 2–4 °C lower than greenhouse air temperature and 4–6 °C lower than the plant temperature of greenhouse without shading net. During night hours also, it was observed that, the plant temperatures were 2–4 °C more than the ambient air temperatures, 1–2 °C less than the greenhouse air temperature and 2–5 °C more than the plant temperatures of greenhouse without shading net. During the night hours the greenhouse air as well as plant temperatures remained with a higher value due to the blanket effect of the shading net whose effects were fairly favorable for the plants like okra during the winter days.

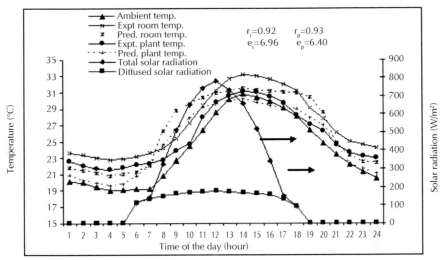

FIGURE 6.11   Hourly variations of plant and greenhouse air temperatures (experimental and predicted), ambient air temperature and solar intensity of greenhouse with shading net on November 5, 2012 from 1 a.m. to 12 midnight.

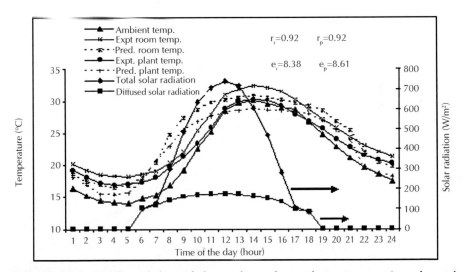

FIGURE 6.12   Hourly variations of plant and greenhouse air temperatures (experimental and predicted), ambient air temperature and solar intensity of greenhouse with shading net on December 15, 2012 from 1 a.m. to 12 midnight.

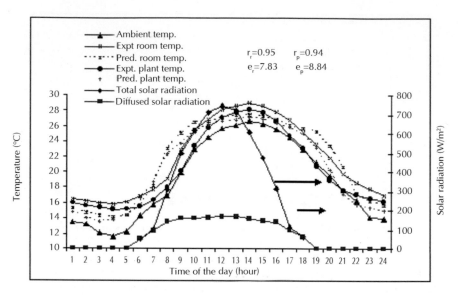

FIGURE 6.13 Hourly variations of plant and greenhouse air temperatures (experimental and predicted), ambient air temperature and solar intensity of greenhouse with shading net on January 12, 2013 from 1 a.m. to 12 midnight.

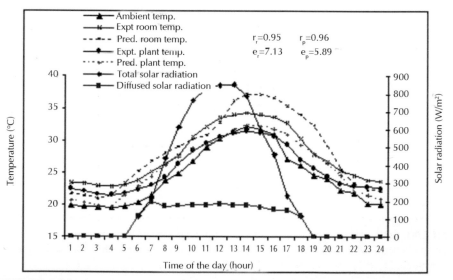

FIGURE 6.14 Hourly variations of plant and greenhouse air temperatures (experimental and predicted), ambient air temperature and solar intensity of greenhouse with shading net on February 4, 2013 from 1 a.m. to 12 midnight.

Greenhouse with shading net increased the greenhouse air temperature up to 2 °C than the ambient air temperatures during peak sunny hours in winter days, which was more favorable for plant growth (okra) than greenhouse without shading net in which the temperatures increased up to 7 °C. Also during night hours in winter days shading net increased temperature up to 5 °C, but in greenhouse without shading net, the temperature increased up to 2 °C compared to the ambient air temperatures.

Also taking into account the peak sunny and mid night hours, the variations of temperature were less in case of greenhouse with shading net than without shading net due to the partial elimination of incoming radiation during sunny hours and prevention of the radiative heat losses to the cold night sky causing better heat distribution inside the greenhouse during night hours. Hence, shading net over the greenhouse is more congenial particularly for the cultivation of okra and cucumber as off-season vegetables during winter days of coastal regions. The vegetable okra can be cultivated three times in a year and its yield was 2.5 times more than that in open field conditions. Similarly the cost of cultivation of okra in an area of 48 sq m was calculated to be Rs. 600 and Rs. 630 both inside greenhouse with shade net and outside condition, respectively.

6.8 CONCLUSIONS

i.   A quasi-steady state thermal model was developed for the greenhouse for predicting greenhouse air temperature and plant temperature. The model in this chapter can be used for studying the performance of greenhouse in a variety of climatic conditions and for various sets of operating conditions.

ii. The increase in air temperatures inside greenhouse without shading net and with shading net over the ambient air temperature was up to 7 °C and 2 °C, respectively, during peak sunny hours; and up to 2 °C and 5 °C, respectively, during night hours in winter days resulting in maintaining the favorable thermal environment of air temperatures in the range of 22–28 °C for greenhouse with shading net for the cultivation of okra.

iii. Natural ventilation was done from 10 a.m. to 4 p.m. to keep the air temperatures inside the greenhouse with shading net within 1–2 °C higher than the ambient air temperature to make it suitable for the growth of the plants okra inside the greenhouse.

iv. The variations of temperature were less in case of greenhouse with shading net than without shading net due to partial elimination of incoming radiation during sunny hours and reducing radiative heat losses to the cold night sky for maintaining better heat distribution inside the greenhouse during night hours due to shading net.

v. The predicted air and plant temperature inside greenhouse without shading net as well as greenhouse with shading net by the model developed were in good agreement with experimental values, which has been confirmed by statistical analysis. Therefore, the proposed model can be used as a design tool for predicting the plant as well as the greenhouse air temperature in the greenhouse, and greenhouse with shading net for heating requirements.

Based on predicted air and plant temperatures inside the greenhouse, crops to be grown inside it can be decided. Also based on the predicted air temperature, heating or cooling requirements for maintaining suitable thermal environment for particular crop inside the greenhouse can also be known to obtain better plant growth and yield.

KEYWORDS

- air temperature
- conduction
- energy equation
- evaporation
- greenhouse
- India, off-season
- okra
- plant temperature
- radiation
- shading net
- solar energy
- sustainable farming
- thermal modeling
- validation
- vegetables
- water use efficiency

REFERENCES

1. Anonymous. *Hand Book of Agriculture*; ICAR Publication: New Delhi, India. 2010; pp 3.
2. Anonymous. *Orissa Agricultural Statistics*; Directorate of Agriculture and Food Production: Government of Odisha, Bhubaneswar. 2012.
3. Chapra, S. C.; Canale, R. P. *Numerical Methods for Engineers*; McGraw Hill Book Co.: New York, 1989.
4. Chen, C.; Shen, T.; Weng, Y. Simple Model to Study the Effect of Temperature on the Greenhouse with Shading Nets. *Afr. J. Biotechnol.* **2011,** *10*(25), 5001–5014.
5. Emekli, N. Y.; Kendirli, B.; Kurunc, A. Structural Analysis and Functional Characteristics of Greenhouses in the Mediterranean Region of Turkey. *Afr. J. Biotechnol.* **2010,** *9,* 3131–3139.
6. Fidaros, D. K.; Baxevanou, C. A.; Bartzanas, T.; Kittas, C. Numerical Simulation of Thermal Behavior of a Ventilated Arc Greenhouse during a Solar Day. *Renew. Energ.* **2010,** *35*(7), 1380–1386.
7. Ganguly, A.; Ghosh, S. Model Development and Experimental Validation of a Floriculture Greenhouse under Natural Ventilation. *Energy Build.* **2009,** *41*(5), 521–527.
8. Goyal, M. R.; Harmsen, E. W. *Evapotranspiration: Principles and Applications for Water Management;* Apple Academic Press Inc.: Oakville, ON, Canada, 2014.
9. Goyal, M. R. *Research Advances for Sustainable Micro Irrigation;* Apple Academic Press Inc.: Oakville, ON, Canada, 2015; Vol. 9.
10. Impron, I. S.; Hemming, G. P. A. Simple Greenhouse Climate Model as a Design Tool for Greenhouse in Tropical Lowland. *J. Biosys. Eng.* **2007,** *69,* 79–89.
11. Kumar, K. S.; Tiwari, K. N.; Jha, M. K. Design and Technology for Greenhouse Cooling in Tropical and Subtropical Regions: A Review. *Energy Build.* **2009,** *41*(12), 1269–1275.
12. Liu, B. Y. H.; Jordan, R. C. Daily Isolation on Surfaces Tilted towards the Equator. *ASHRAE J.* **1962,** *3*(10), 53.
13. Omer, A. M. Construction, Applications and the Environment of Greenhouses. *Afr. J. Biotechnol.* **2009,** *8,* 7205–7227.
14. Oz, H.; Atilgan, A.; Buyuktas, K.; Alagoz, T. The Efficiency of Fan-Pad Cooling System in Greenhouse and Building up of Internal Greenhouse Temperature Map. *Afr. J. Biotechnol.* **2009,** *8,* 5436–5444.
15. Panwar, N. L.; Kaushik, S. C.; Kothari, S. Solar Greenhouse an Option for Renewable and Sustainable Farming. *Renew. Sustain. Energ. Rev.* **2011,** *15*(8), 3934–3945.
16. Sengar, S. H.; Kothari, S. Thermal Modeling and Performance Evaluation of Arch Shape Greenhouse for Nursery Raising. *Afr. J. Math. Comput. Sci. Res.* **2008,** *1*(1), 1–9.
17. Sethi, V. P.; Sharma, S. K. Survey of Cooling Technologies for Worldwide Agricultural Greenhouse Applications. *Sol. Energ.* **2007,** *81,* 1447–1459.
18. Singh, G.; Singh, P.; Singh, P. P.; Lubana, K. G. Formulation and Validation of a Mathematical Model of the Microclimate of a Greenhouse. *Renew. Energ.* **2006,** *31*(10), 1541–1560.
19. Singh, R. D.; Tiwari, G. N. Energy Conservation in the Greenhouse System: A Steady State Analysis. *Energy.* **2010,** *35*(6), 2367–2373.

APPENDIX – I. LIST OF SYMBOLS IN THIS CHAPTER.

A Area (m²)

C Specific heat (J/kg °C)

F_n Fraction of solar radiation falling on glazed north wall, dimensionless (decimal)

F_p Fraction of solar radiation falling on plants in greenhouse, dimensionless (decimal)

h_i Heat transfer coefficient from greenhouse cover to inside greenhouse (W/m² °C)

h_0 Heat transfer coefficient from greenhouse cover to the ambient environment (W/m² °C)

h_p Convective and evaporative heat transfer coefficient from plant mass to greenhouse air (W/m² °C)

h_{cp} Convective heat transfer coefficient from plant mass to greenhouse air (W/m² °C)

h_{ew} Evaporative heat transfer coefficient from water mass to greenhouse air (W/m² °C)

h_{gr} Heat transfer coefficient from floor to the air in greenhouse (W/m² °C)

$h_{g\infty}$ Heat transfer coefficient from floor of greenhouse to higher depth of ground (W/m² °C)

h_{gca} Convective heat transfer coefficient from greenhouse cover to the ambient air (W/m² °C)

h_{pgc} Convective heat transfer coefficient from plant in greenhouse to greenhouse cover

$h_{r,\,gca}$ Radiative heat transfer coefficient from greenhouse cover to the ambient air (W/m² °C)

$h_{r,\,pg}$ Radiative heat transfer coefficient from plant to greenhouse (W/m² °C)

K Thermal conductivity (W/m °C)

K_s Thermal conductivity of shade cloth (W/m °C)

L_g Thickness of ground (m)

K_s Thickness of shade cloth (W/m °C)

M_a Mass of air in greenhouse (kg)

M_p Mass of plants in greenhouse (kg)

N Number of air changes per hour in greenhouse

$P(T)$ Saturated vapor pressure at temperature, $T(P_a)$

R_n Reflectivity of glazed north wall from inner side of greenhouse, dimensionless, decimal

T Temperature (°C)

t Time in seconds

U	Overall heat loss coefficient from greenhouse air to the ambient air through greenhouse covers (W/m² °C)
V	Volume of greenhouse (m³)
v	Velocity of air (m/s)

SUBSCRIPT

a	Ambient air
g	Floor of greenhouse
i	Number of walls and roofs of greenhouse
n	Glazed north wall of greenhouse
p	Plant
S	Shade cloth
r	Greenhouse (Greenhouse enclosure)
gc	Greenhouse cover
eff	Effective

GREEK LETTERS

α	Absorptivity (decimal)
ρ	Density (kg/m³)
σ	Stefan-Boltzmann constant (5.67×10^{-8} W/m² K⁴)
φ	Emissivity, dimensionless
τ	Transmissivity of greenhouse cover, dimensionless
τ_s	Transmissivity of shade cloth, dimensionless
λ	Latent heat of vaporization (J/kg)
γ	Relative humidity (decimal)

CHAPTER 7

USE OF GREEN ENERGY SOURCES FOR MICRO IRRIGATION SYSTEMS

M. K. GHOSAL

Department of Farm Machinery and Power, College of Agricultural Engineering and Technology, Orissa University of Agriculture and Technology, Bhubaneswar 751003, Odisha, India

E-mail: mkghosal1@rdiffmail.com

CONTENTS

ABSTRACT

The growing demands of energy and water particularly in the agricultural sector have necessitated the adoption of reliable, environment-friendly, and water-saving technologies so as to combat energy crisis and drought in near future. It has been established that conventional sources of energy (oil, gas, coal, etc.) will not be able to provide desired levels of energy security to mankind in foreseeable future. Hence, there is a global consensus for exploitation and utilization of different sources of energy which are reliable, naturally available, and environment-friendly. These sources of energy are otherwise known as green energy because of protecting environment, conserving natural resources, and increasing sustainability.

Development of conventional forms of energy at a reasonable cost is the responsibility of the government. However, limited fossil fuel resources and associated environmental problems have emphasized the need for new and sustainable energy supply options. We are still heavily dependent on coal and oil for meeting our energy demand, which contribute to smog, acid rain, and greenhouse gases' emission. Hence, popularization and approach for switching over slowly from conventional to non-conventional sources of energy (green energy) should be the present day's strategy of the society so far as the economic feasibility, social desirability, and environmental soundness are concerned.

In this chapter, the effective use of green energy sources particularly in the agricultural activities has been discussed briefly in order to provide the appropriate insights among the farming community to adopt those highly reliable, non-polluting, and naturally available energy sources for overall sustainable development.

7.1   INTRODUCTION

The consumption of energy has been almost increasing exponentially since the industrial revolution. To meet the growing demand of energy, there is at present the pressing need of alternative sources of energy in order to provide solutions to the present-day problems with the fossil fuels. The target is then to explore such energy systems that have no negative environmental, economic, and societal impacts, which we mostly refer to as "green energy." The sources of green energy include sun, wind, biomass, geothermal, hydropower system, etc., which will provide an important attribute for sustainable development [10]. The practical relevance of deriving power from the sun especially through photovoltaic (PV) system is now gaining momentum in the tropical countries because of profuse and year round availability of clear sun shine and the decreasing cost of solar modules day by day.

In view of this, this chapter discusses the applicability of green energy including solar energy in the agricultural sector for sustainable development and environmental protection.

Energy has become the prime commodity in modern civilization and the amount of energy consumption has become the indicator for the standard of living of a nation. It has long been recognized that the excessive use of energy has the adverse impact on the environment, economy, and society, from local air and water pollution to the threat of global warming (the mean temperature increase around the globe) and climate variability (the temperature fluctuations around the mean); and from the economic difficulties arising out of the rapid increase and swings in energy prices [8]. The sustainable development of humanity and the economy with the security of energy has at present topped national agendas around the world. It is imperative to develop energy strategies, policies, and technologies to achieve this objective through an energy system(s) that have no observable (or net) negative impact on environment, economy, and society. At present, most of the energy requirement worldwide is met by the combustion of fossil fuels (i.e., coal, petroleum oils, natural gas, etc.) [5], which have become an essential and integral part of modern civilization, being increasingly relied upon since the industrial revolution. Only a very small proportion of the energy comes from nuclear and hydro power, and a much smaller portion from solar, wind, hydro, geothermal, tidal wave, and so on. This almost exclusive reliance on the combustion of fossil fuels has resulted in enormous amounts of harmful pollutant emissions to our environment, and has caused severe degradation of the local and global environment, and has exposed the world population (from humans to animals and from plants to all forms of life on earth) to the hazards and risks created by the extensive use of fossil fuels [8, 12].

In addition to the health and environmental concerns, a steady depletion of the world's limited fossil fuel reserves also calls for alternative primary energy sources and new energy technologies for energy conversion and power generation that are more energy efficient than the conventional combustion engine with minimal or no pollutant emissions and add to sustainable development [4]. Key requirements for sustainable development include societal, economic, and environmental sustainability, all related to the sustainability of energy systems by reducing the use of only dependant fossil fuels. Alternatives to using fossil fuels include use of large-scale PV devices, solar thermal system, intensive biomass cultivation, wind turbines, and small scale hydroelectric projects, etc. For catering the growing needs of the society and because of the problems that are being faced due to the excessive usage of the fossil fuels and due to their scarcity, the demand for the alternative sources of energy (green energy or sustainable energy) is growing day by day.

7.2 CONVENTIONAL AND RENEWABLE SOURCES OF ENERGY

Improvements in quality of life and rapid industrialization in many countries are increasing energy demand significantly, and the potential future gap between energy supply and demand is predicted to be large. Interest in sustainable development and growth has also grown in recent years, motivating the development of environmental benign energy technologies. Research on applications of solar energy technologies has as a consequence expanded rapidly, exploiting the abundant, free, and environmentally characteristics of solar energy [4]. However, widespread acceptance of solar energy technology depends on efficiency, cost-effectiveness, reliability, and availability. Renewable energy sources can be defined as "energy obtained from the continuous or repetitive currents on energy recurring in the natural environment" or as "energy flows which are replenished at the same rate as they are used." All the earth's renewable energy sources are generated from solar radiation, which can be converted directly or indirectly to energy using various technologies. This radiation is perceived as white light since it spans over a wide spectrum of wavelengths, from the short-wave infrared to ultraviolet. Such radiation plays a major role in generating electricity either producing high temperature heat to power an engine mechanical energy which in turn drives an electrical generator or by directly converting it to electricity by means of the PV effect.

The PV is the simplest technology to design and install, however, it is still one of the most expensive renewable technologies. It is environmentally friendly and a non-pollutant low maintenance energy source. Nature needed about 300 million years to create hard coal or crude oil from the organic matter of plants but mankind is consuming it much faster and at the same time blowing CO_2 into atmosphere, creating greenhouse effect, global warming, and climate changes. At present, the energy mix is composed mainly of coal, natural gas, and crude oil as well as nuclear power with the advantage of guaranteed supply and continuity but also with known disadvantages of CO_2 emissions associated with fossil fuels. Renewable energy sources like solar, biomass, and wind are now therefore gaining importance for energy security, reliability, and economic development.

7.3 CONCEPT OF GREEN TECHNOLOGY

The green technology is a broad term for more environmentally friendly solutions. Green technology can be used as environmental healing technology that reduces environmental damages created by the products and technologies for peoples' conveniences. It is believed that green technology promises to augment farm profitability while reducing environmental degradation and conserving natural resources. The risk associated with "dirty" technologies such as the petroleum products are alarmingly raising. The "clean" technologies are expected to provide low risk alternatives. Green technology covers a broad group of methods and materials for generating energy to non-toxic cleaning products. The reason for this approach has been significantly important as people expect a dramatic innovation and changes in their livelihood. The development of alternative technology should attempt to benefit the planet by truly protecting the environment. Although it is difficult to precisely define the areas that are covered by green technology, yet green technology helps in addressing the emerging issues of sustainability because of the advancement in science and technology. This technology should meet the needs of society in ways that can continue indefinitely into the future without damaging or depleting natural resources. In short, green technology is defined as the technology that meets present needs without compromising the ability of future generations to meet their own needs. In terms of the technology, products can be fully reclaimed or re-used by successfully reducing waste and pollution in the cradle to grave cycle of manufacturing process. The innovations in technology have aroused interest in developing alternative fuels as a new means of generating energy and energy efficiency. The use of such type of technology is also very much important in the agricultural sector to achieve food security for the fast growing population and mitigating the present-day's concern of global warming and climate change.

Irrigation is the most energy consuming activity in farming system. Irrigation of small-holdings is likely to become increasingly important and widely used in the next decades, especially in developing countries, because of increasing population pressure and because the majority of land-holdings are small, particularly in Asia and Africa. Hence, the most viable options for deriving power through green energy technology are from solar, biomass, and wind energy. PV pumping systems are particularly well-suited for water- and energy-saving methods of irrigation such as drip and sprinkler irrigation because of their advantage in the low-energy range. Ever increasing fuel price and unreliable electricity supply are at present the major constraints for achieving assured irrigation for crop cultivation. Solar pump may therefore be an alternative to the electric motor operated pumps for irrigated crop production especially for the off-grid rural areas.

Being an environmentally sound and green technology, solar pumps can be promoted in the agricultural sector [3]. Electricity produced from PV systems has a far smaller impact on the environment than traditional methods of electrical generation. During their operation, PV cells need no fuel, give off no atmospheric or water pollutants, and require no cooling water. Unlike fossil fuel (coal, oil, and natural gas) fired power plants, PV systems do not contribute to global warming or acid rain. The use of PV systems is not constrained by material or land shortages and the sun is a virtually endless energy source. The cost of PV systems has decreased more than twenty times since the early 1970's, and research continues on several different technologies in an effort to reduce costs to levels acceptable for wide scale use. Current PV cells are reliable and already cost-effective in certain applications such as remote power, with stand-alone PV plants built in regions not reached by the utility networks.

Pande et al. [7] designed and developed a solar PV operated pump drip irrigation system for growing orchards in arid region considering different design parameters like pump size, water requirements, the diurnal variation in the pressure of the pump due to change in irradiance and pressure compensation in the drippers. Meah et al. [6] discussed some policies to make solar photovoltaic water pumping (SPVWP) system an appropriate technology for the respective application region as it has proved its aspects technically, economically, and environmentally in developed countries. Short et al. (2003) investigated some of the issues involved in solar water pumping projects, described the positive and negative effects that they can have on the community and proposed an entirely new type of pump, considering the steps that could be taken to ensure future sustainability. Badescu (2003) analyzed the operation of a complex time dependent solar water pumping system consisting of four basic units: a PV array, a battery, a DC motor, and a centrifugal pump.

7.4   SOLAR PHOTOVOLTAIC SYSTEM

Electrical energy is the pivot of all developmental efforts in the developed and the developing nations because conventional energy sources are finite and fast depleting. In the last decades, energy related problems are becoming more and more important and involve the ideal use of resources, the environmental impact due to the emission of pollutants and the consumption of conventional energy resources. Direct solar energy conversion to electricity is conventionally done using PV cells, which makes use of the PV effect [1]. PV effect depends on interaction of photons, with energy equal to, or more than the band-gap of PV materials. Some of the losses due to the band-gap limitations are avoided by cascading semiconductors of different band-gaps. PV modules generate electricity directly from light without emissions, noise, or vibration. Sunlight is free but power generation cost is exceptionally high, although prices are starting to come down. Solar energy has low energy density as PV modules require a large surface area for small amounts of energy generation.

The primary component in grid connected PV systems is the inverter that converts DC power produced by PV array into AC power consistent with the voltage and power quality requirement of the utility gird. Silicon solar cells are perhaps the simplest and most widely used for space and terrestrial applications. The PV system is promising source of electricity generation for energy resource saving and CO_2 emission reduction, even if current technologies are applied. Further the development in efficiency of solar cells, amount of material used in the solar cell, and the system design for maximum use of recycled material will reduce the energy requirement and greenhouse gas emissions. Net annual CO_2 emission mitigation potential from 1.8 kW solar photovoltaic pump at an average solar radiation of 5.5 kWh/m^2/day is about 2085 kg from diesel operated pumps and about 1860 kg from petrol operated pumps. The CO_2 emissions mitigation potential is higher in the case of diesel substitution as compared to the petrol substitution. This is primarily due to low efficiency of fuel utilization in the diesel engine pump. The demand of electricity in irrigation is growing up, since the cost of an electric powered pump is lower compared to a diesel engine driven pump. Solar pump may be an alternative for small scale irrigated crop production in the off-grid areas of the tropical countries endowed with abundant supply of solar energy.

The solar radiation varies from 4.0 to 6.5 kWh/m^2/day and the bright sunshine hours vary from 6 to 9 h/day. There is a vast area to be irrigated where most of the areas have no grid connection. Solar PV pumps can be used for irrigating these lands for better crop production and to increase cropping intensity. In regions with high insolation levels, PV pumping systems were technically suitable for use, beneficial for the environment and were cheaper over the diesel engine driven pumps [1]. The energy crisis is severe during the irrigation season, which is a threat to economical development of any country. Using solar pumps on a large scale, energy demand in irrigation systems can be reduced substantially. Though the initial cost of a solar pump is higher than a conventional diesel engine operated pump, solar pump has lower maintenance cost, which makes it cost-effective over the years. Moreover, a solar pump is pollution free and environment friendly water pumping system.

7.5   SOLAR WATER PUMPING: A CASE STUDY [9]

Electrical and diesel powered water pumping systems are now-a-days widely used for irrigation applications. The continuous exhaustion of conventional energy sources and their environmental impacts have created an interest in choosing solar photo-voltaic pumping system in a sustainable manner. Farmers are still using hand pumps for irrigating small patch of land. Electric and diesel operated pump sets are mostly used for lifting water from dug well, bore well, and other irrigation systems. Lifting of water by hand pump is a most tedious and labor consuming operation. Similarly, nonavailability and erratic supply of grid connected electricity in the remote areas and rising cost of diesel day by day necessitate the search of a reliable source of energy for assured irrigation. Installation of electric pump sets is not at all possible at most of the locations as the agricultural fields are far away from the electric grid station. In addition, the electric tariff is increasing in every year and thus increasing the cost of water pumping operation. Further, the repair and maintenance cost of electric motor operated pump sets is generally higher than that of solar PV water pumping system. When not much research work was conducted on solar PV water pumping system, then diesel pumping system was very popular among the farming community due to its low cost and portability. During this time, the diesel cost was also cheaper. But it causes environmental pollution and global warming by releasing a considerable amount of CO_2 into the atmosphere. The repair and maintenance cost of diesel pump set is also higher than that of solar PV water pumping system. Hence, solar PV water pumping system is today a superior option left for the farming community as its pumping cost is cheaper as compared to electric and diesel pump sets. Moreover, the risk of environmental pollution is less and its repair and maintenance cost is very low. It can be installed at any location as per the desire of the farmers, because solar energy is available profusely and free of cost in the nature.

Sprinkler irrigation is getting popular due to its numerous advantages over other surface irrigation methods. In this case study, summer groundnut crop (cv. TG-22) was cultivated at central research station, Orissa University of Agriculture and Technology (OUAT), Bhubaneswar during 2009–2010 using sprinkler irrigation in one plot and conventional furrow irrigation in the other. Comparison of sprinkler irrigation method with that of conventional furrow irrigation method included water use efficiency (WUE), water productivity, saving of labor, and yield potential. It was observed that sprinkler irrigation system has 20% higher WUE, 32% saving in water, 22% saving in labor, and 20% increase in yield over the conventional furrow irrigation method. The experimental set-up is shown in Figures 7.1 and 7.2.

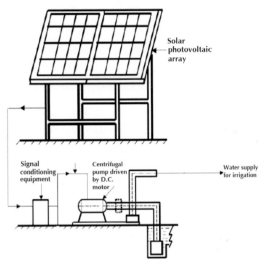

FIGURE 7.1   Solar PV system installed at the field site.

FIGURE 7.2   SPVWP system.

7.5.1   LOCATION

Experiment was carried out at OUAT central research station, Bhubaneswar in a sandy soil. The soil has 90.6% sand, 2.0% silt, 7.4%clay with bulk density of 1.59 g/cm^3 and pH of 5.52. Two plots, each measuring 60 × 60 m^2 were taken where sprinkler and conventional furrow irrigation methods were practiced to grow summer groundnut (var TG-22). Each plot was again divided into three subplots, where different depths of irrigation were applied. Volume of water supplied to each subplot was measured. Yields from these plots were recorded.

7.5.1.1   DETERMINATION OF ET$_0$

Daily reference crop evapotranspiration (ET$_0$) was computed by taking climatological data into consideration by using modified Penman method. The crop coefficient curve for summer groundnut was developed following the procedures outlined in the guidelines for predicting crop water requirements [9] to determine evapotranspiration of the crop for the growing period.

7.5.1.2   DETERMINATION OF CROP COEFFICIENT

The effect of climate was taken into consideration while computing ET$_0$. But the effect of crop characteristics was not incorporated; and hence, to study the effect of crop characteristics, the crop coefficient was determined from a prepared crop coefficient curve. Crop stage period with crop coefficient (K_c) for groundnut crop are given in Table 7.1.

TABLE 7.1   Duration of Different Growth Stages and Crop Coefficient (K_c) of Groundnut (TG-22).

Crop variety	Duration in days			
	Initial stage	Crop development stage	Mid-season stage	Late season stage
Groundnut TG-22	Dec 22 to Jan 15	Jan 16 to Feb 14	Feb 15 to Mar 31	April 1 to April 19
K_c	0.26	0.26– 1.0	1.0	0.58

The ET$_{crop}$ was calculated as follows:

$$ET_{crop} = K_c \times ET_0 \qquad (7.1)$$

where: K_c = Crop coefficient; and ET$_0$ = Reference crop evapotranspiration in mm/day.

7.5.1.3 UNIFORMITY COEFFICIENT

The uniformity coefficient was computed using sprinkler distribution test by Christiansen and was 85%.

7.5.1.4 WATER PRODUCTION FUNCTION

Crop production functions were developed using the empirical relationships between crop yield and the various input parameters. The actual moisture used by the crop during the irrigation interval was gravimetrically determined by measuring the soil moisture up to a depth of 80 cm in the root zone for both irrigation methods. Table 7.2 shows data on depth of water application, amount of nitrogen, crop yield, and WUE for sprinkler and conventional furrow irrigation methods, respectively. Multivariate production functions were developed using these data.

TABLE 7.2   Input Output Data for Groundnut Grown Under Two Methods of Irrigation.

Types of irrigation	Treatment	Input (water applied), mm	Nitrogen applied, kg/ha	Output (groundnut yield), (100 kg)/ha	WUE, kg/ha-mm
Sprinkler	S$_1$	502	20.0	27.30	5.43
	S$_2$	537	40.0	28.23	5.25
	S$_3$	572	60.0	28.67	5.02
Furrow	F$_1$	606	20.0	22.33	3.68
	F$_2$	653	40.0	24.00	3.66
	F$_3$	700	60.0	24.00	3.42

The general production function was expressed in the form of:

$$y_c = a_0 + a_1 W_c + a_2 N + a_3 W_c^n + a_4 N^n + a_5 W_c^n \qquad (7.2)$$

$$y_c = f \left(\frac{W_c \times N}{P} \right)$$ (7.3)

$$y_c = aW_c bN^d$$ (7.4)

where: y_c = Pod yield in (100 kg)/ha; W_c = Irrigation water applied in mm; N = Nitrogen level in kg/ha; a_0 = Combine effect of all the fixed inputs; a_1, a_2, a_3, a_4, and a_5 are regression coefficients and n is the exponent that defines degree of a polynomial; P = Quantity of other fixed inputs that denote that only W_c and N are variable inputs; and a, b, c are coefficients.

Eq 7.3 was fitted to data for y_c, W_c, and N with exponent values of $n = 1, 2,$ 1.5, and 0.5 to obtain linear, quadratic, three-halves, and square root production functions. The data were also fitted to the Cobb–Douglas function or power function in Eq 7.4. Of all the combinations tried, a linear production function was most satisfactory and was given in the following form:

$$Y_{gs} = a_c + a_1 W_{gs}$$ (7.5)

$$Y_{gf} = a_c' + a_1' W_{gf}$$ (7.6)

where: Y_{gs} = Pod yield of groundnut in sprinkler method in (100 kg)/ha; W_{gs} = Water applied to groundnut through sprinkler method in mm; Y_{gf} = Pod yield of groundnut in furrow irrigation in (100 kg)/ha; a_c and a_1 are regression coefficients; and W_{gf} = Water applied under furrow irrigation method in mm.

It is evident that application of nitrogen fertilizer did not influence the pod yield of groundnut. Therefore, the effect of nitrogen on pod yield was neglected and only the effect of irrigation water on the pod yield was considered. Thus, the linear relationship with only water as a variable appears to be justified in a situation, in which the crop has grown. In conventional furrow irrigation method, water was applied to the plot according to the local practices. Soil samples were taken after 24 h of irrigation from the depths of 20, 40, and 60 cm to determine the soil moisture percentage by gravimetric method. Yield potential of groundnut was recorded by taking 10 samples from 1 m² area from both the plots. It is one of the parameters, which was used for computation of WUE. WUE is the yield of marketable crop per unit depth of water used in evapotranspiration:

$$WUE = \frac{Y}{ET}$$ (7.7)

where Y = Marketable yield of crop in kg/ha; ET = Evapotranspiration of the crop in mm/day; and WUE = Water use efficiency in kg/ha-mm.

7.5.1.5   FIELD WUE (E_U)

Field WUE is the ratio of the crop yield to the amount of water used in the field (WR), which includes $(G + ET + D)$.

$$F_u = \frac{Y}{G + ET + D} = \frac{Y}{WR} \tag{7.8}$$

where, G = Water used for metabolic purpose of the crop in mm (negligible); WR = Water used in the field in mm; and D = Deep percolation in mm.

In this section, the depth of water needed for proper growth of the crop was determined. The irrigation factor was taken into consideration keeping all other input parameters constant. Yields with respect to water requirement in both the methods were determined.

7.5.2   RESULTS FOR GROUNDNUT

7.5.2.1   CROP EVAPOTRANSPIRATION

The weekly crop evapotranspiration for the growing period of summer groundnut is presented in Figure7.3.

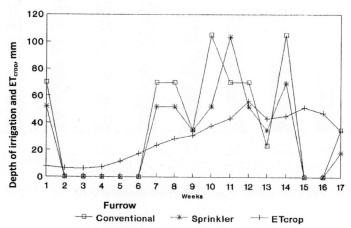

FIGURE 7.3   Weekly ET$_{crop}$ and depth of irrigation under sprinkler and furrow methods.

It can be observed from Figure 7.3 that crop evapotranspiration was maximum during the 12th week of the growing season that is, toward the end of the mid season stage. At this stage, the depth of water applied by the sprinkler method was almost equal to the crop evapotranspiration. During 12th week, depth of water under sprinkler is 52 mm, whereas, the crop evapotranspiration is 56 mm, which are approximately equal. After 12th week, weekly crop evapotranspiration was decreased gradually. Also during the first five weeks, water applied under furrow and sprinkler methods was much higher than the ET_{crop}. This was because soil needed more water for germination of seeds. Again during this period, the soil temperature was very low because of mild winter, which elongated the germination period. During the entire crop season, 653 mm of water was applied in furrow method and 537 mm in sprinkler method, whereas the total crop water requirement during the cropping season was 503 mm. Thus, water supplied in sprinkler method was optimum. In furrow method of irrigation, 22% excess water was applied as compared to under sprinkler method.

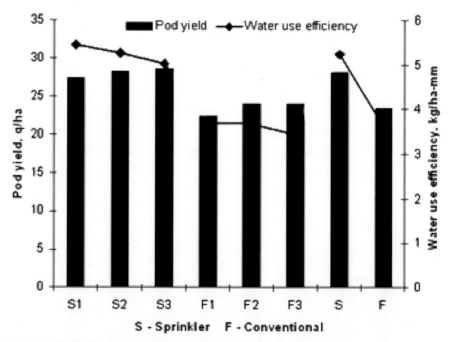

FIGURE 7.4   Pod yield and WUE in sprinkler and furrow method.

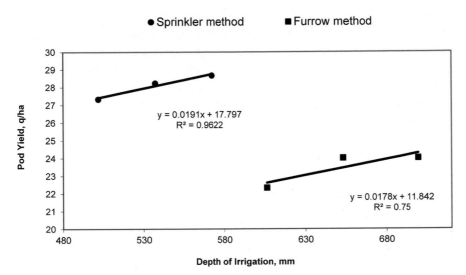

FIGURE 7.5   Water production functions of summer groundnut in sprinkler and furrow methods.

7.5.2.2 YIELD POTENTIAL

In sprinkler method, the average yield was 2790 kg/ha compared to 2330 kg/ha (Figs. 7.4 and 7.5). An increase of 20% yield in sprinkler method was observed over furrow irrigation method. Weights of pods (1000 pods) were 1620 and 1220 gm, respectively, under sprinkler and furrow irrigated plots indicating a better quality of kernel in case of sprinkler irrigation.

7.5.2.3   WUE

Consumptive WUEs were 5.58 and 4.65 kg/ha-mm in sprinkler and furrow methods, respectively (Fig. 7.4). It was observed that WUE in sprinkler method was 20% higher than under conventional furrow method.

7.5.2.4   FIELD WUE

Average field WUEs were 5.23 and 3.56 kg/ha-mm in sprinkler and furrow methods, respectively. The increase in application of water from 502 to 572 mm in sprinkler did not show any appreciable rise in yield as the WUE declined from 5.43 to 5.02 kg/ha-mm (Table 7.2). In sprinkler method, consumptive WUE and field WUE were almost same for application of 502 mm depth of water. Hence, 502 mm of water application in sprinkler method was considered optimum. Water requirement per 100 kg of produce under sprinkler method was 19.11 ha-mm, which indicated a saving of 32% of water compared to conventional furrow method.

7.5.2.5   WATER PRODUCTION FUNCTION

For sprinkler method of irrigation:

$$Y_{gs} = 17.80 + 0.019\ W_{gs} \tag{7.9}$$

For conventional furrow method of irrigation:

$$Y_{gf} = 11.84 + 0.017\ W_{gf} \tag{7.10}$$

From Eqs. (9) and (10), it is evident that application of nitrogen fertilizer did not influence the pod yield of groundnut. This is because groundnut requires only 4% nitrogen to support the plant in early stage and being a leguminous crop, it is capable of fixing nitrogen from atmosphere by symbiosis and meets its requirement after 20 days after seeding. Thus linear relationship with only water as a variable was justified. From these equations and Figure 7.5, it was inferred that pod yield is directly proportional to the quantum of water used. The increase in yield was 1.9 and 1.7 kg per mm of water used under sprinkler and furrow methods, respectively.

7.5.2.6   ECONOMICS OF SPRINKLER AND FURROW METHOD OF IRRIGATION FOR SUMMER GROUNDNUT

While comparing the labor requirements, a 22% saving was observed in sprinkler method over furrow method. Water and labor savings could have been enhanced if a large field was taken up for commercial purpose. The return per rupee of investment, as shown in Table 7.3, was Rs.3.73 in furrow method of irrigation as compared to Rs.5.49 in sprinkler method of irrigation. The return was 47.18 % higher in sprinkler method in comparison to conventional furrow method of irrigation.

TABLE 7.3 Cost–Benefit Ratio for Two Irrigation Methods.

Method of irrigation	Cost of cultivation, Rs/ha	Total return, Rs/ha	Net profit, Rs	Cost–benefit Ratio
Sprinkler	9168	59,590	50,422	5.49
Furrow	10,133	48,930	38,597	3.73

The following conclusions were drawn from this case study for groundnut:

a. Sprinkler irrigation method has significantly higher WUE compared to conventional furrow method of irrigation.
b. There is a significant saving of labor in sprinkler method over conventional furrow method. Saving of labor is higher where land leveling is needed.
c. Sprinkler irrigation method is advantageous for groundnut production or any other close growing crops where water is a limiting factor.
d. Initial investment is higher in sprinkler irrigation method but it can be compensated in the long run.
e. There is significant saving of water and increase in pod yield of groundnut. The yield potential can be enhanced further when a large field area is considered.
f. In most part of the growing season, the water applied through sprinkler method was higher than that of ET_{crop} and was optimum.
g. From the water production functions, it was concluded that the increase in yield per unit depth of water was higher in sprinkler method compared to conventional furrow method of irrigation.

7.6 CASE STUDY: DRIP WATERING SYSTEM IN ZERO ENERGY COOL CHAMBER (ZECC) [2]

A considerable amount of perishable horticultural produce is wasted every year due to lack of appropriate storage facilities [11]. In tropical climatic conditions, maintenance of low temperature is a great problem [2]. Mechanical cooling is energy intensive, expensive, and not easy to install and run in rural areas. The zero energy cool chamber (ZECC), utilizing the principle of evaporative cooling, is reported to maintain relatively low temperature and high humidity compared to ambient conditions which is required for short term storage of fruits and vegetables. Evaporative cooled storage structures are designed to reduce air temperature in cooling applications through the process of evaporation of water. Now-a-days much emphasis is being given in the cultivation of horticultural crops like fruits and vegetables as they are highly remunerative and grown in a short period compared to cereal crops. The wide variation in the environmental conditions poses huge difficulty in storing fresh fruits and vegetables. The majority of farmers are usually small and marginal categories and has poor resource availability. In the absence of proper storage technique, the farmers usually sell the vegetables in the local markets soon after the harvest. This situation very often compels for a distress sale of the products at very low price. ZECC with drip irrigation system through gravity flow are becoming more effective for safe storage of vegetables. Drip irrigation system is used for uniform wetting of sand layer for proper evaporation to occur resulting into decrease in temperature and increase of humidity in ZECC. Application of water in the chamber plays a vital role in regulating temperature and RH. Too dry cool chamber will not provide the desired cooling effect and too moist chamber causes unnecessary wastage of water and may sometimes lead to fungus growth. Therefore, it is necessary to find out the optimum quantity of water needed under different situations of seasonal variations to achieve effective performance of the chamber.

An experiment was conducted during the year 2015 to evaluate the efficacy of ZECC on the storability of leafy vegetables (amaranthus, spinach, and coriander). These leafy vegetables were grown in 50 m² (10 × 5m²) area each in the central farm of OUAT from February to April 2015. The ZECC (Figs. 7.6, 7.7, and 7.8) was constructed in the premises of College of Agricultural Engineering and Technology, OUAT, Bhubaneswar, which is coming under warm and humid climatic region and in the coastal belt of Odisha. The quality of the stored leafy vegetables in ZECC was studied during peak summer from April 20, 2015 to April 28, 2015 when there was high temperature with low relative humidity outside.

ZECC is an on-farm rural oriented storage structure, which operates on the principle of evaporative cooling and it was constructed using locally available raw materials such as bricks, sand, bamboo, rice straw, vetiver grass, jute cloth, etc. The chamber was constructed above ground and comprised of a double-walled structure made up of bricks. The cavity of the double wall was filled with riverbed sand. The lid was made by using vetiver grass mat on a bamboo frame. Floor of the chamber was made with the help of bricks and of size 165 × 115 cm². The space between the double wall was 7.5 cm, filled with fine sand. A protective cover for the inner chamber was made with the help of a bamboo frame and closely spaced dry vetiver grass. A thatched shed over the chamber was erected to protect it from direct sun or rain. After construction of the chamber, the following procedure was adopted for its use:

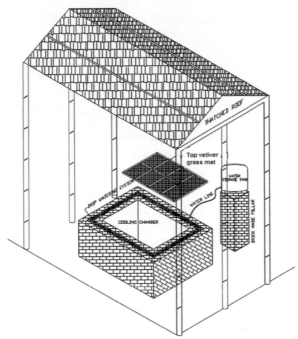

FIGURE 7.6 A typical ZECC.

FIGURE 7.7 ZECC with micro-dripper system under gravity flow.

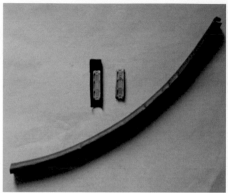

FIGURE 7.8   Micro dripper.

a. The walls of the chamber were made wet and the sand in between the double wall was saturated with water by the gravity fed micro-dripper from a 35 L capacity plastic bucket placed 50 cm from the top of the chamber (Figs. 7.7 and 7.8).

b. Fresh leafy vegetables to be stored were weighed and taken in perforated bags.

c. Polythene bag placed inside the chamber.

d. The chamber was closed completely with the wetted vetiver grass mat.

e. The walls of the chamber as well as the sand filled in the gap of the double wall structure were watered twice daily (morning and afternoon) to maintain high relative humidity and low temperature inside the chamber.

f. All the three leafy vegetables (amaranthus, spinach, and coriander) were kept at a time in ZECC to study their storability during peak summer period.

g. The temperature and relative humidity inside the chamber were monitored periodically using a hand held portable RH/temperature meter. Simultaneously, a control sample in each of the leafy vegetable was also maintained under room conditions to compare the effectiveness of ZECC.

7.6.1 QUALITATIVE EVALUATION

7.6.1.1 DETERMINATION OF THE PHYSIOLOGICAL LOSS IN WEIGHT (PLW)

PLW is one of the main factors to determine the quality of stored vegetables. Observations of PLW and the shelf-life of greens were monitored every day using a digital electronic balance and Eq 7.11. The readings were made at one day interval during the experiment period. The shelf-life of vegetables was determined on the basis of 5–10% PLW. A decrease of PLW up to 10% often results in a loss of freshness and wilted appearance.

$$\text{Physiological loss in weight, percentage} = \frac{(X1 - X)}{X} \times 100 \qquad (7.11)$$

where, $X1 =$ Initial weight in g; $X =$ Weight in g, at the end of storage time. The following procedure was adopted:

a. The physiological loss in weight of the produce under storage was measured on daily basis.
b. The weights of rotten and unmarketable vegetables either due to development of fiber or pithiness were also measured during storage on daily basis.
c. The temperature and the relative humidity inside the chamber, room condition, and their deviation from ambient were recorded at periodical intervals on the days of storage using a hand held portable RH/temperature meter.

7.6.1.2 EXPERIMENTAL FINDINGS

The loss in weight of the stored leafy vegetables (PLW) is an indication of moisture loss from the produce, which renders the leafy vegetables unmarketable as they lose freshness. The moisture loss from horticultural produce is not a mere loss of weight rather it is a loss of appearance, taste, and even nutrients from the produce which ultimately results in economic loss of the produce. So, any storage method for perishables like leafy vegetables should aim at minimizing the moisture loss and respiration from the produce so as to enhance the quality and marketability. This is possible by reducing the storage temperature and increasing the relative humidity of air surrounding the produce in the storage atmosphere. The salient findings of the study are as follows (Tables 7.4 and 7.5):

a. Decrease in temperature in the range of 3–7 °C and increase in RH by 12–26% inside the ZECC compared to room conditions were recorded at various hours of a day during the experimental observations. Similarly, decrease in temperature in the range of 6–10 °C and increase in RH by 25–40% inside the ZECC compared to outside ambient conditions were recorded at various hours of a day during the experimental period. The highest differences were mostly observed at 2.00 pm.

b. The PLW was recorded for amaranthus on day 7 of storage inside the ZECC, and it was only 10.85% as against 20.12% under room conditions. The rotten percentages were 12.56 on day 7 and 31.54 on day 2 in ZECC and room conditions, respectively. Considering the PLW (10.85%) and rotten/unmarketable greens (12.56%) together, amaranthus can be well stored in the ZECC for seven days. Similarly, allowing PLW (20.12%) and rotten/unmarketable greens (31.54%) together, amaranthus can be well stored under room conditions only for two days.

c. The PLW for spinach on day 4 of storage inside the ZECC was only 11.64% as against 40.38% under room conditions. The rotten percentages were 20.54 on day 4 and 36.78 on day 2 in ZECC and room conditions, respectively. Considering the PLW (11.64%) and rotten/unmarketable greens (20.54%) together, spinach can be well stored in the ZECC for four days. Similarly, allowing PLW (14.97%) and rotten/unmarketable greens (28.87%) together after one day, spinach can be well stored under room conditions only for one day.

d. The PLW for coriander on day 2 of storage inside the ZECC was only 10.32% as against 44.78% under room conditions. The rotten percentages were 10.98 on day 2 and 30.56 on day 1 in ZECC and room conditions, respectively. Considering the PLW (10.32%) and rotten/unmarketable greens (10.98%) together, spinach can be well stored in the ZECC for two days. Similarly, allowing PLW (10.56%) and rotten/unmarketable greens (30.56%) together after one day, coriander can be well stored in the room condition only for one day.

e. The garden bean was found to have a storability of three days in ZECC (PLW of 9.21 and 8.37% of unmarketable pod), whereas radish can be well stored for a week with a PLW of around 11% without the loss of freshness.

TABLE 7.4 Experimental Temperature and Humidity for Storage of Selected Leafy Vegetables in ZECC and in Room Condition during Summer Period.

Leafy vegetable	Recommended temperature and humidity and shelf-life (day)			Experimental temp., humidity and shelf-life (day) in ZECC			Experimental temp., humidity and shelf-life (day) in room condition		
	Temp. (°C)	RH (%)	Shelf-life (day)	Temp. (°C)	RH (%)	Shelf-life (day)	Temp. (°C)	RH (%)	Shelf-life (day)
Amaranthus	5–8	90–95	10–14	20–33	80–95	7	24–37	60–78	2
Spinach	5–8	90–95	8–10	20–33	80–95	4	24–37	60–78	1
Coriander	5–8	90–95	6–10	20–33	80–95	2	24–37	60–78	1

TABLE 7.5 Shelf-Life of Some Leafy Vegetables in ZECC and Room Condition during Summer Period.

Leafy vegetable	Shelf-life in ZECC (day)	Shelf-life in room condition (day)
Amaranthus	7	2
Spinach	4	1
Coriander	2	1

One of the major constraints faced by marginal and small farmers engaged in cultivation of leafy vegetables is the perishability of the produce which forces them to sell the produce to whatever prices offered in the nearby market point. Taking the produce to a distant market is not feasible owing to the small quantity of sundry leafy vegetables being harvested every day. On the other hand, the consumers are also paying a high price for a poor quality produce as there is no cold chain market facilities established in rural and sub urban areas. Considering these constraints, the farmers growing multiple leafy vegetables on a small scale, some alternate technologies for storing their produce at least for a shorter period will help so that the everyday harvest can be accumulated and taking a bigger volume of the produce once in a few days to the market becomes an economically viable option for the grower. The consumer is also benefited by the availability of fresh and nutritious leafy vegetables.

7.7 BIOGAS TECHNOLOGY

Energy is an indispensable part of modern society and can serve as one of the most important indicators of socio–economic development. Despite advancements in technology, however, majority of people, primarily in the rural areas still continue to meet their energy needs for cooking through traditional means by burning biomass resources (i.e., firewood, crop residues, and animal dung) in the most inefficient conventional cook stoves. Such practices are known to be the source of significant environmental, social, economic, and public health issues. To achieve sustainable development in these regions, it is imperative that access to clean and affordable (renewable) energy is made available. In this context, upgrading existing biomass resources (i.e., animal manure, crop residues, kitchen waste, and green waste) to cleaner and more efficient energy carriers (such as biogas from anaerobic digestion (AD)) has unique potential to provide clean and reliable energy, while simultaneously preserving the local and global environment. Hence, the use of existing biomass such as kitchen waste, cattle dung, crop residues, green waste, and the organic fraction of industrial and municipal wastes for producing clean and renewable energy through AD both in rural and urban areas would improve human health, the local environment, and the socio–economic conditions. AD is a biological process that converts organic matter into energy-rich biogas in the absence of oxygen. Biogas, a mixture primarily consisting of CH_4 and CO_2 can be used as a clean renewable energy source for cooking, generating heat, and electricity, and can be upgraded into biomethane for use as a transportation fuel as well. Biogas digestate, a nutrient-rich residue following digestion, can be used as a soil conditioner and/or organic fertilizer. Thus, AD through biogas technology can play a significant role in addressing all of the aforementioned concerns of effective waste management and reliable energy generation along with simultaneously increasing agricultural productivity.

7.8   BIODIESEL: THE ALTERNATE, VEGETABLE-BASED FUEL

Biodiesel is a cleaner-burning diesel fuel made from natural, renewable sources such as vegetable oils. Just like petroleum diesel, biodiesel can be used in combustion-ignition engines. Biodiesel is made through a chemical process called trans-esterification, whereby the glycerin is removed from the vegetable oil. It is produced from soybean or other vegetable oil or from used cooking oil (there are more than four billion gallons of waste cooking oil produced annually in the USA, enough to replace 10% of fuel expenditures). It can be made from almost any plant-derived oil. It decreases solid carbon fraction of particulate matter (since the oxygen in biodiesel enables more complete combustion to CO_2), eliminates the sulphate fraction (as there is no sulfur in the fuel). Therefore biodiesel works well with new technologies such as catalysts which reduce the soluble fraction of diesel particulate but not the solid carbon fraction (Table 7.6).

- Biodiesel reduces carbon monoxide, carbon dioxide, sulfur dioxide (one of the main causes of acid rain), hydrocarbons, benzene, and particulate matter.
- Increases NO_x levels (unless biodiesel is made from used cooking oil). The biodiesel industry is looking for additives that would reduce NOx levels.
- Can safely be blended with petroleum diesel.
- Can be used in any diesel engine with little or no modification to the engine or the fuel system.
- Results in a slight drop in fuel economy.
- Can extend the life of diesel engines.
- Is biodegradable.
- Would create new jobs and extra income for farmers.
- Provides a domestic renewable energy supply.
- Is safer to use than petroleum diesel: It has a flash point of 300 °F (vs. 125 °F for diesel).
- Can also be used as a fuel lubricity additive in diesel fuel.

TABLE 7.6   Comparative Studies of Some Green Energy Fuels.

Fuel type	Engine issues	Storage issues	Availability	Environment impact
CNG	No modification to engine required	Large storage tank minimizes luggage capacity	Available mainly in Delhi and Mumbai regions	25% lower CO_2 emission as compared to petrol and diesel
LPG	No modification to engine required	Storage tank is smaller but still reduces luggage capacity	Available in around 619 stations mainly in Southern and Western India	15% lower CO_2 and 50% lower particulate emissions as compared to petrol. 10% lower CO_2 and 90% lower particulates as compared to diesel
Biodiesel	Minor modification required if the blend is > 25%	No issues as it goes into the normal fuel tank	Large scale crop is required but can be grown in waste land	For B100 fuel, 30% lower particulate emissions and 80% lower CO_2 for life cycle emissions as compared to diesel
Ethanol	Minor modification required if the blend is > 24%	No issues in storage	Depends on sugarcane crop and the processing capacity	E100 fuel can reduce net CO_2 emissions up to 100% on a life-cycle basis and 22–50% on usage basis

7.9 CONCLUSIONS

There is no doubt that green energy sources would play critical role in ensuring energy security of a country. There is enormous potential to generate energy from these sources like solar and biomass. The government in most of the countries has also been very actively involved in promoting green energy. It is high time for the government to develop a comprehensive green energy policy and design support schemes for the benefit of the farming community. The exploitation of green energy gives agriculture the opportunity to migrate from the fossil-based net energy consumer to the ecologically friendly net energy generator. The ability to yield electricity from green energy mostly from solar energy presents land farms with alternative sources of income, long-term employment opportunities are being created, not to mention a healthy infrastructure and stabilization of population growth in rural areas but also giving due importance and priority to the agriculture sector. It supports decentralized self-sufficient energy production and creates employment in other sectors such as trade, industry, and services and reduces dependency on fossil fuel, which is finite and depleting fast. Green energy in the form of sunshine, wind, flowing river streams, and biomass has traditionally been used as major sources of energy for carrying out various domestic, community, and agricultural production activities. Since 1960s, commercial sources of energy like electricity, diesel, petrol, etc. have gradually replaced the traditional sources of energy to a large extent. As of now, whole world is struggling with the problem of shortage of grid power.

Small and medium industries, services and agricultural sectors are now facing the consequences of inadequate power supply. Out of these, rural sector has been the worst sufferer. The situation is unlikely to improve in near future. Besides, increase in prices of crude oil in international market have compelled most of the countries to vigorously pursue measures for conservation of the petroleum fuels by increasing their use efficiency and also develop their suitable substitutes using indigenously available resources. Keeping this in view, it has now become imperative that the traditionally used green energy sources need to be exploited efficiently, particularly in the countryside where these are available in abundance. During the last two decades, fairly good number of green energy technologies have been developed and commercialized in the country for rural applications. Still there is a necessity of wide popularization of these technologies for the benefit of the farming community.

KEYWORDS

- **biofuels**
- **coal**
- **cost–benefit ratio**
- **cost economics**
- **economic feasibility**
- **evapotranspiration**
- **furrow irrigation**
- **gas**
- **gas emission**
- **green energy**
- **groundnut**
- **India**
- **micro irrigation**
- **natural resources**
- **oil**
- **solar energy**
- **solar pump**
- **sprinkler irrigation**
- **sustainable farming**
- **vegetables**
- **water use efficiency**
- **yield potential**

REFERENCES

1. Abu-Aligah, M. Design of Photovoltaic Water Pumping System and Compare It with Diesel Powered Pump. *Jordan J. Mech. Ind. Eng.* **2011,** *5*(3), 273–280.
2. Anonymous. *A Study on Short Term Storage of Leafy Vegetables (Greens) in Zero Energy Cool Chamber*; Unpublished B. Tech (Agri. Eng.) Project Report: CAET, OUAT, Bhubaneswar, Odisha, India, 2015.
3. Biswas, H.; Faisal, H. Solar Pump: A Possible Solution of Irrigation and Electric Power Crisis of Bangladesh. *Int. J. Comput. Appl.* **2013,** *62*(16), 1–5.

4. Demirbas, F. M. Biofuel for Sustainable Development. *Appl. Energy.* **2011,** *88,* 3473–3480.
5. IEA, *World Energy Outlook*; International Energy Agency (IEA): Paris, France, 2011.
6. Meah, K.; Ula, S.; Barrett, S. Solar Photovoltaic Water Pumping Opportunities and Challenges. *Renew. Sust. Energ. Rev.* **2008,** *12,* 1162–1175.
7. Pande, P. C.; Singh, A. K.; Ansari, S.; Vyas, S. K.; Dave, B. K. Design Development and Testing of a Solar PV Pump Based Drip System for Orchards. *Renew. Energ.* **2003,** *28,* 385–396.
8. Parida, B.; Iniyan, S.; Goic, R. A Review of Solar Photovoltaic Technologies. *Renew. Sust. Energ. Rev.* **2011,** *15,* 1625–1636.
9. Sahoo, N.; Ghosal, M. K. In a *Comparative Study of Sprinkler and Furrow Irrigation Methods in Summer Groundnut*, Souvenir 7th National Seminar on Emerging Climate Change Issues and Sustainable Management Strategies: OUAT, Bhubaneswar, Odisha, India, 8–10 February, 2014.
10. Singh, J.; Sai, G. Biomass Conversion to Energy in India-A Critique. *Renew. Sust. Energ. Rev.* **2010,** *14,* 1367–1378.
11. Tayde, P. Creating Wealth from Organic Waste. *Akshay Urja.* **2012,** *5*(6)*,* 38–40.
12. Tiwari, G. N.; Ghosal, M. K. *Fundamentals of Renewable Energy*; Narosa Publishing House: New Delhi, India, 2007.

PART III
Emerging Technologies

CHAPTER 8

NUTRIENT MANAGEMENT THROUGH DRIP FERTIGATION TO IMPROVE THE YIELD AND QUALITY OF MULBERRY CROP

K. ARUNADEVI[1], P. K. SELVARAJ[2], and V. KUMAR[3]

[1]Department of Soil and Water Conservation Engineering, Agricultural Engineering College and Research Institute, Kumulur 621712, Tamil Nadu, India

[2]Agricultural Research Station, TNAU, Bhavanisagar 638451, Tamil Nadu, India

[3]Department of Agricultural Engineering, Agricultural College and Research Institute, Tamil Nadu Agricultural University, Madurai 625104, Tamil Nadu, India

CONTENTS

ABSTRACT

Water is the most limiting factor in the Indian agricultural scenario. Further unscientific use of the available irrigation water compounds the problems in crop production. It is becoming increasingly clear that with the advent of the high yielding varieties, the next major advance in our agricultural production is expected to come only through efficient water management technologies.

Of all the inputs in mulberry cultivation, irrigation is the most important to optimize leaf productivity and determine its growth, development, and all metabolic activities. Cultivation of mulberry plant is mainly for the commercial production of silk. Most of the irrigation is by open system having a relatively low efficiency of water application. It is therefore, important to adopt right scheduling of irrigation for achieving maximum water use efficiency. Maximization of crop yield and quality and minimization of leaching below the rooting volume may be achieved by managing fertilizer concentrations in measured quantities of irrigation water, according to crop requirements.

There is lack of information on drip irrigation and fertigation with normal fertilizers for different crop geometries and drip layouts in mulberry crop. Hence, a study was conducted to arrive a suitable economic drip irrigation layout with efficient and optimum irrigation scheduling of mulberry crop through drip irrigation and to optimize the fertilizers nitrogen and potassium through fertigation.

Irrigation and fertigation levels influenced plant height, number of branches per plant, number of leaves per branch, leaf area, and leaf area index. Maximum plant height, number of branches per plant, number of leaves per branch, leaf area, and leaf area index were observed under single row drip irrigation at 80% of surface irrigation level followed by paired row drip and micro-tube irrigations at 80% of surface irrigation level.

8.1   INTRODUCTION

India's crop production suffers mainly from the availability of water, and water is the most limiting factor in the Indian agricultural scenario. Further unscientific use of the available irrigation water compounds the problems in crop production. In the present era of acute water shortage, caused by over utilization and depletion of both surface and subterranean water resources, employment of suitable water management practices for effective utilization of available resources in an economic way is of prime importance. Rapid increase in mulberry production in India during the last century could be achieved through irrigation. It is necessary to develop a proper irrigation scheduling for mulberry and to optimize the water and fertilizer requirement under drip fertigation with different crop geometries. Water, N and K fertilizer requirements should be optimized.

Mulberry is a hardy perennial plant and it is the main food plant for the silkworm. Cultivation of mulberry plant is mainly for the commercial production of silk. Next to China, India is the largest producer of silk in the world. All the known varieties of silk, viz., Mulberry, Eri, Muga, and Tassar are produced in India. Mulberry is the most popular variety in India, contributing more than 87% of the country's silk production. Mulberry belongs to the genus *Morus* of the family *Moraceae*. It is capable of thriving under a variety of agro-climatic conditions. Propagation is generally done through vegetative methods. Leaf yield in mulberry is the only required produce. The quality of the mulberry leaf can be decided by its contents of moisture, protein, carbohydrates, and fiber. They vary with several factors like soil texture, moisture content, variety, planting system, agronomical practices, etc.,

Through the planned progress made in mulberry cultivation, raw silk production has increased enormously to the planned target level (Table 8.1). However, the quality of raw silk produced in India is poor. Even the best quality silk does not come anywhere near the "A grade" silk of International Standards. For producing good quality silk, good quality of mulberry leaves should be given as feed for cocoon production.

TABLE 8.1 Mulberry Silk Production in India.

Years	Mulberry plantation ('000 ha)	Reeling cocoon ('000 tons)	Raw silk (tons)	Silk waste (tons)	Renditta
1980–81	170	58	4593	1376	12.70
1981–82	180	55	4801	1523	11.50
1982–83	197	67	5214	1825	12.80
1983–84	207	71	5681	2017	12.50
1984–85	215	75	6895	2464	10.90
1985–86	218	77	7029	2504	10.90
1986–87	230	82	7905	2837	10.30
1987–88	242	87	8455	3086	10.20
1988–89	268	96	9683	3399	10.00
1989–90	294	110	10,805	3921	10.10
1990–91	317	117	11,486	3953	10.20

TABLE 8.1   *(Continued)*

Years	Mulberry planta-tion ('000 ha)	Reeling cocoon ('000 tons)	Raw silk (tons)	Silk waste (tons)	Renditta
1991–92	331	107	10,658	3544	10.00
1992–93	343	130	13,000	4513	9.98
1993–94	319	117	12,550	4518	9.34
1994–95	283	123	13,450	4501	9.15
1995–96	286	116	12,884	4022	9.03
1996–97	281	132	12,954	4000	8.93
1997–98	282	127	14,048	4250	9.08
1998–99	270	127	14,260	4250	8.88
1999–00	227	125	13,994	4153	8.96
2000–01	216	125	14,432	4237	8.66
2001–02	232	140	15,842	4655	8.81
2002–03	194	128	14,617	4514	8.77
2003–04	185	117	13,970	3764	8.41
2004–05	172	120	14,620	3587	8.21
2005–06	179	126	15,445	3749	8.17
2006–07	192	135	16,525	4055	8.20
2007–08	185	135	16,245	3416	8.12
2008–09	178	125	15,610	3746	8.00
2009–10	184	132	16,322	4080	8.07
2010–11	170	131	16,360	4090	7.99
2011–12	181	140	18,272	4568	7.66

Compiled by: Economic Division, Updated on September 27, 2012
Source: Central Silk Board, India, http://ministryoftextiles.gov.in/ermiu/Mulberry_Silk_Production.pdf

It is estimated that for the preparation of a silk saree weighing 400 g, 6000 cocoons, which consume as much as 1.5 million mulberry leaves and requiring 100 man-hours at rearing stage of cocoons are required. Since there is practically no scope for further increase in the net sown area, the only alternative is to raise the productivity per unit area, per unit of input especially water and fertilizer, and per unit of time.

Developing countries like India need to boost the economic status of poor farmers through adoption of new technologies to increase crop production through the optimal use of scarce resources, such as land, water, and fertilizers. Major reasons for low productivity levels in India are lack of adequate irrigation facilities and unbalanced fertilizer scheduling.

Of all the inputs in mulberry cultivation, irrigation is the most important to optimize leaf productivity and determine its growth, development and all metabolic activities. Per capita availability for agricultural purposes is shrinking every year due to depletion of ground water. Application of manure and fertilizers are also required for increasing productivity and improved quality of mulberry leaves. Failure of monsoons and uncontrolled exploitation of ground water calls for added importance of the efficient utilization of available water and therefore improved water management becomes mandatory.

As water is becoming increasingly scarce, adoption of micro irrigation system offers potential for bringing nearly double the area under irrigation with the same quantity of water. Drip irrigation had saved irrigation water to the tune of 40–70% coupled with yield increase as high as 100% in some crops in some specific locations [4].

Drip irrigation and fertigation have gained enormous popularity, owing to a significant saving in water and fertilizer use compared to the conventional methods. Tamil Nadu farmers are following surface irrigation to mulberry crop. Surface irrigation methods lead to problems like erosion, waterlogging, evaporation; deep percolation and leaching of fertilizers thereby decrease in the crop yield, and cocoon production. Fertigation is a modern technology, wherein water and soluble fertilizers are applied simultaneously in a combined form to the soil root zone resulting in minimal loss of nutrients and water. Hence, modern irrigation methods like drip, micro sprinkler, etc., are to be adopted along with fertigation to save water and fertilizers besides improving the yield and quality of the produce.

There is no definite information about exact quantity of water required and scheduling of irrigation to mulberry with drip irrigation. Hence, there is a need for an intensive study to determine the water requirement and scheduling of irrigation for different crop geometries and layouts to facilitate cost reduction and increased efficiency of the drip system without sacrificing the plant population and yield in a given area.

High-grade water-soluble fertilizers and liquid fertilizers are costly and for our farmers, use of these costly fertilizers is impracticable. Among the normal fertilizers mostly used by farmers, urea for nitrogen and muriate of potash for potassium are water soluble. If these fertilizers are used for fertigation, there will be reduction in fertilizer cost and farmers will come forward for fertigation.

This study was conducted to fix up optimum irrigation schedule through drip irrigation for mulberry and to explore the possibility of increasing the productivity, fertilizer use efficiency (FUE) and water use efficiency in mulberry with drip irrigation, and fertigation.

8.2   PRACTICES IN MULBERRY CULTIVATION

It is essential to select mulberry variety according to agro-climatic conditions of the area. The varieties selected should have good agro-economic efficiency in terms of their response to applied fertilizers or FUE, disease resistance, and drought tolerance. The leaves of a selected variety should have good succulence, shelf life, palatability, and nutritive value. The test crop variety was Victory-1 (V_1), the new mulberry variety evolved by Central Sericultural Research and Training Institute, Mysore, holds high potentials under irrigated conditions in South India. This superior mulberry variety is suitable to different agro-climatic conditions that not only yield better, but also will be of high quality to support the growth of silkworm and resistant to climatic hazards, diseases and insects. It is suitable for semi-arid region.

Cultivation of mulberry plant is mainly for the commercial production of silk. Rapid increase in mulberry silk production in India during the last century could be achieved through irrigation. As water and labor have become limiting factors and too expensive, which necessitate the use of most efficient method of irrigation at farm level. Most of the irrigation is by open system having a relatively low efficiency of water application. It is therefore, important to adopt right schedule of irrigation for achieving maximum water use efficiency. So an efficient irrigation management system is needed by which production of the mulberry is enhanced even at low irrigation water input. Drip irrigation generates a restricted but concentrated root system requiring frequent nutrient supply that may be satisfied by applying fertilizers in irrigation water, that is, by fertigation. Maximization of crop yield and quality and minimization of leaching below the rooting volume may be achieved by managing fertilizer concentrations in measured quantities of irrigation water, according to crop requirements.

Hence, fertilizer and irrigation are the two factors, which play major role in increasing mulberry output.

This study was conducted to optimize irrigation scheduling through drip irrigation for mulberry and to explore the possibility of increasing the productivity, FUE and water use efficiency in mulberry with drip irrigation, and fertigation.

8.2.1 FLOOD IRRIGATION IN MULBERRY

In surface irrigation methods, there are more leaching loss of nutrients, insufficient moisture availability due to long irrigation intervals, and soil compaction.

Flooding and irrigation with small furrows were the age old and traditional systems in our country, which bear very low irrigation efficiencies, especially the distribution uniformity. The conventional methods of irrigation had poor irrigation efficiency as low as 25–30%.

The drawbacks of the flooding type of irrigation system in mulberry crop included non-uniform application of water, impounding in certain pockets, loss of water due to percolation, and leaching of nutrients due to excess water application.

The water source through bore well or open well was becoming dry during the summer season as a result of which the farmers were either forced to cut short the brushing quantity or stop the rearing of silkworm during the season.

8.2.2 WATER MANAGEMENT IN MULBERRY

Development of efficient water management system is very essential for economizing the water use. The problem of water management can be solved by determine the water requirement that is, when and how much to irrigate and find out the adoption of irrigation for different soil and adopt an efficient method of irrigation.

In drip irrigation methods, the moisture content is maintained at field capacity level leading to luxurious growth and increased yield of mulberry.

8.2.3 DRIP IRRIGATION IN MULBERRY CROP

Micro irrigation or drip irrigation as is popularly known is an ingenious method of irrigation, wherein water and soluble nutrients are delivered near the plants.

Due to advancement in designing and reduced cost of material, drip irrigation is becoming more popular and acceptable for crops like mulberry.

Drip irrigation could help immensely in enhancing the quality and freshness of mulberry leaves, as it was possible to supply precise quantity of water to the plants just before harvesting of the leaves.

Subba Rao et al. [10] reported that mulberry leaf yield in the field level was only 43.9% (6586 kg vs. 15,000 kg ha^{-1} y^{-1}) under rainfed conditions and 47.4% (16,590 kg vs. 35,000 kg ha^{-1} y^{-1}) under irrigated conditions.

Studies conducted by the Central Sericultural Research and Training Institute (CSRTI), Mysore on water management in mulberry revealed that a minimum of 40% of irrigation water could be saved by drip system over the surface irrigation. Besides higher leaf moisture content, higher values of nitrogen, phosphorus, and potash contents, increase in leaf yield was also recorded under drip irrigation [8].

Drip system was found more efficient than furrow system resulting in increased mulberry leaf yield by 10.3–14.5% [2].

It has been observed by other investigators that rainfall in the arid and semi-arid zones is very erratic and prevailing drought conditions adversely affect mulberry leaf production. The essential requirement of irrigation is to provide the appropriate amount of water to the mulberry crop at right time through uniform distribution system so that water is not wasted at all.

The production function analysis indicated that fertilizer and irrigation were the two factors, which played major role in increasing mulberry output in Salem district in Tamil Nadu [7].

8.2.4   NUTRIENT MANAGEMENT IN MULBERRY

The improvement in the yield and quality of mulberry can be achieved through application of fertilizers. Out of 16 essential nutrients, nitrogen is the first major nutrient, which needs greater attentions. Mulberry absorbs nitrogen either as the ammonia (NH_4^+)) or the nitrate or the nitrate ion (NO_3^-). Broadcasting of fertilizers is not advisable in mulberry garden because it causes loss on Nitrogen from the soil through volatilization. Some nitrogen sources, such as urea, ammonium sulphate, etc., may lose ammonia by volatilization as a result of improper placement. This can be corrected by drip fertigation method.

Potassium is a major plant nutrient because of large amount, in which it is absorbed by plants and its significant role on high yield and quality. In mulberry, potassium plays an important role in various biochemical functions, development, and yield of foliage in addition to the improvement in the leaf quality.

In order to get the optimum yield of quality mulberry leaves, it is essential that nitrogen should be applied in optimum doses in mulberry cultivation along with phosphorus and potassium. A balanced nutrient supply should be ascertained for healthy growth and maximum biomass (i.e., leaf yield) production in mulberry cultivation.

Application of high-grade water soluble fertilizers particularly under drip system can save the cost on fertilizers by 20–30% and by 30–40% on water. Fertigation in mulberry has various advantages like, higher use efficiency of both water and fertilizer, minimum losses of nitrogen due to prevention of leaching, optimization of nutrient balance of N, P_2O_5, and K_2O by supplying these nutrients directly to the root zone of the crop, regulation, and monitoring of fertilizer doses through desired nutrient concentration in solution to effect timely supply of nutrients on a continual basis and saving in application cost and improvement in physical and biological conditions of soil.

Thus, drip fertigation can help immensely in enhancing the quality and freshness of mulberry leaves as it is possible to supply the precise quantity of water and fertilizer to plants.

8.2.5 FERTIGATION IN MULBERRY

In fertigation through drip, the nutrient losses can be minimized to a considerable extent, if fertilizers are applied in the effective crop root zone in required quantities and by maintaining optimum soil moisture regime.

With drip irrigation, there was often an intimacy between roots and applied water. So it was feasible through fertigation, to manipulate the nutrition of drip-irrigated plants [6].

Fertigation permitted application of fertilizer formulations directly to the active root site and thus improved the nutrient use efficiency. Reduction of 20–25% in fertilizer dose was reported [9].

Maximization of crop yield and quality and minimization of leaching below the rooting volume might be achieved by managing fertilizers concentrations in measured quantities of irrigation water, according to crop requirements [3].

Micro irrigation offered potential for fertigation. Through fertigation, accurate and uniform application was possible and the amounts and concentrations of specific nutrients could be adjusted to crop requirements [1].

Jeyabal et al. [5] reported that fertigation was recognized as a very effective and convenient means of maintaining optimal fertility and water supply.

8.3   METHODS AND MATERIALS

A study was conducted to arrive a suitable economic drip irrigation layout with efficient and optimum irrigation scheduling in mulberry crop through drip irrigation and to optimize the fertigation of nitrogen and potassium.

Maximum plant height, number of branches per plant, number of leaves per branch, leaf area, and leaf area index were observed under single row drip irrigation at 80% of surface irrigation followed by paired row drip and micro-tube irrigations at 80% of surface irrigation level.

8.4   RESULTS AND DISCUSSION

Single row drip at 80% of surface irrigation registered more leaf yield followed by paired row drip and micro-tube at 80% of surface irrigation level than in micro irrigation at 60 and 40% of surface irrigation level. Among the various drip layouts, single row drip irrigation system was found excellent compared to other layouts (paired row drip and micro-tube).

Drip irrigation levels had exerted favorable influence on leaf quality parameters when compared to surface irrigation. Maximum coarse leaf moisture content (60.09%), tender leaf moisture content (72.76%), leaf nitrogen content (3.97%), leaf potassium content (2.13%), coarse leaf protein content (15.98%), and tender leaf protein content (24.30%) were observed under single row drip irrigation at 80% of surface irrigation level.

Fertigation of N and K fertilizers at 100% of recommended dose exerted significant influence on leaf growth, yield, and quality parameters when compared to conventional fertilizer application by band placement.

In drip irrigation treatments, a considerable saving in irrigation water was observed. Drip treatments resulted in irrigation water saving of 14.66, 29.51, and 44.56% at 80, 60, and 40% of surface irrigation levels, respectively, compared to surface irrigation treatment. The mean quantum of irrigation water use under surface irrigation for experimental season was 503.40 mm, whereas in drip treatments it were 429.59, 354.86, and 279.07 mm, respectively, at 80, 60, and 40% of surface irrigation level. Among the different irrigation and fertigation levels, single row drip at 80% of surface irrigation with 100% of recommended fertilizer dose was the best treatment with the highest water use efficiency of 27.24 kg ha^{-1} mm^{-1}. The least water use efficiency of 11.59 kg ha^{-1} mm^{-1} was observed in surface irrigation treatment with 100% of recommended fertilizer dose.

The highest Nitrogen use efficiency of 159.40 kg ha^{-1} kg of N^{-1} and Potassium use efficiency of 426.95 kg ha^{-1} kg of K^{-1} were recorded in single row drip irrigation at 80% of surface irrigation with 75% of recommended fertilizer level. The least N and K fertilizer use efficiencies were recorded in surface irrigation treatment (62.16 kg ha^{-1} kg of N^{-1} and 166.50 kg ha^{-1} kg of K^{-1}, respectively) with 100% of recommended fertilizer level.

Benefit-cost (BC) ratio was significantly influenced by both irrigation and fertigation levels. Irrigating mulberry through micro irrigation at 80% of surface irrigation registered the maximum BC ratio for various drip layouts followed by micro irrigation at 60% and 40% of surface irrigation level. BC ratio differed critically with the levels of fertigation application. Application of fertigation at 75% of the recommended level recorded the highest BC ratio compared to 100% of recommended fertilizer level. The maximum BC ratio recorded in micro-tube irrigation was 3.09 followed by paired row drip (2.93) and single row drip (2.56) at 80% of surface irrigation level with 75% of recommended fertilizer level.

Among various drip layouts, single row drip irrigation layout at different levels of surface irrigation registered favorable growth parameters, leaf yield, leaf quality, water use efficiency, and FUE than paired row drip and micro-tube irrigation.

For getting increased yield (nearly 100%), drip irrigation at 80% of surface irrigation and fertigation at 100% of recommended N and K was the best combination with a water saving of 14.66%. In moderately water scarcity areas, drip irrigation at 60% of surface irrigation and 75% of recommended N and K through fertigation could be adopted to get nearly 55% increased yield with water saving of 29.51% and fertilizer saving of 25% in N and K. In severely water scarce locations, drip irrigation at 40% of surface irrigation and 75% of recommended N and K through fertigation was the best choice to get 15% increase in yield least water use (water saving = 44.56%) compared to surface irrigation with 5 cm depth at weekly intervals besides 25% saving in N and K fertilizers. Considering the economics, paired row layout with micro-tubes followed by paired row drip layout at 80% of surface irrigation level with 75% recommended fertilizer dose were the best.

8.5   CONCLUSIONS

A properly designed and well-maintained drip irrigation system for mulberry cultivation can minimize the wastage of water, conserve the soil, reduce the labor costs and brings down power consumption with increase in yield, and improve quality. The water and power thus saved can be used to bring in additional areas under drip irrigation for increasing the acreage of mulberry product. Irrigation with drip method produced significantly higher irrigation water use efficiency compared to furrow irrigation and was due to higher yield under drip irrigation. The leaves obtained from fertigated plots were of high quality than the leaves from locally fertilized and irrigated plots.

KEYWORDS

- **drip irrigation**
- **fertigation**
- **fertilizer use efficiency**
- **India**
- **mulberry cultivation**
- **nitrogen use efficiency**
- **potassium use efficiency**
- **silk**
- **water requirement**
- **water use efficiency**

REFERENCES

1. Bar-Yosef, B.; Sagiv, B. Response of Tomato to N and Water Applied via Trickle Irrigation System. *Agron. J.* **1982,** *74,* 633–638.
2. Benchamin, K. V.; Syed Nizamuddin; Sabitha, M. G.; Asis Ghosh. Mulberry Cultivation Techniques under Water Stress Condition. *Indian Silk.* **1997,** *36*(3)*,* 12–18.
3. Hagin, J.; Lowengart, A. Fertigation for Minimizing Environmental Pollution by Fertilizers. *Fert. Res.* **1995/1996,** *43*(1/3), 5–7.
4. INCID, *Annual Report;* Indian National Committee on Irrigation and Drainage: New Delhi, 1994.

5. Jeyabal, A.; Sundar Raman, S.; Muralidhar Rao, M. ; Palaniappan, S. P.; Chelliah, S. Water Soluble Fertilizers: A Potential Source of Nutrients for Foliar Spray and Fertigation. *Agro India*. **2000,** *4,* 34–36.
6. Keng, J. C. W.; Scott, T. W.; Lugo-Lopez, M. A. Fertilizer Management with Drip Irrigation in an Oxisol. *Agron. J.* **1979,** *71,* 971–980.
7. Lakshmanan, S.; Geethadevi, R. G. An Economic Analysis of Factors Influencing Mulberry Leaf Production in Tamil Nadu. *Indian J. Seric.* **2002,** *41*(2), 120–123.
8. Mishra, R. K.; Choudhury, P. C.; Das, P. K.; Ghosh, A. Sustainable Techniques for Mulberry Cultivation. *Indian Silk.* **1996,** *34*(11), 7–10.
9. Phene, C. J.; Fouss, J. L.; Sanders, C. C. Water, Nutrient, Herbicide Management of Potatoes with Trickle Irrigation. *Am. Potato J.* **1979,** *56,* 51–59.
10. Subba Rao, M.; Shivakumar, G. R.; Magadum, S. B.; Datta, R. K. Yield Gap Studies in Tamil Nadu. *Indian Silk.* **1997,** *36*(4), 16–18.

APPENDIX – I   MULBERRY PRODUCTION IN INDIA.

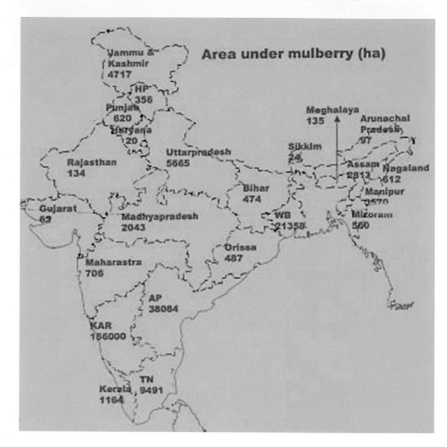

APPEINDIX – II   USE OF FARM MACHINES IN MULBERRY CULTIVATION.

APPENDIX – III MULBERRY CULTIVATION IN INDIA.

Pair row

APPENDIX – IV   FLOW DIAGRAM FOR SILK PRODUCTION.

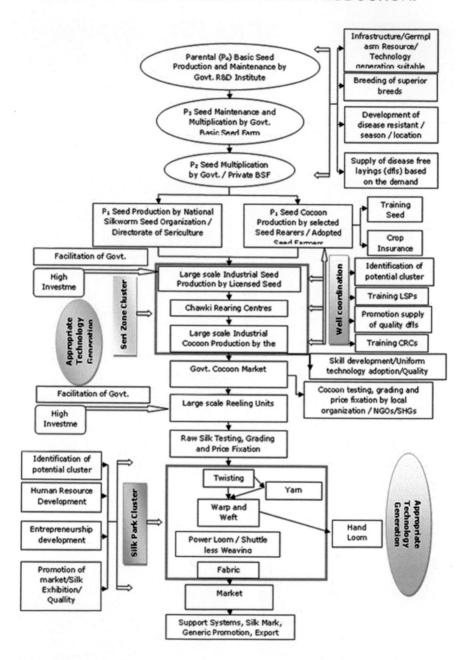

APPENDIX – V   PUBLICATION ON MULBERRY CULTIVATION.

Arunadevi K.

Drip Irrigation and Fertigation on Mulberry Cultivation

Irrigation Scheduling, Water saving and Cost economics

LAP LAMBERT
Academic Publishing

CHAPTER 9

WATER USE EFFICIENCY FOR BELL PEPPER (*CAPSICUM*) UNDER GREENHOUSE CULTIVATION

RANJAN KUMAR DAS and M. K. GHOSAL

Department of Farm Machinery and Power, College of Agricultural Engineering and Technology, Orissa University of Agriculture and Technology, Bhubaneswar 751003, Odisha, India

CONTENTS

ABSTRACT

The suitability of a low-tech naturally ventilated greenhouse was evaluated for off-season cultivation of vegetables in warm and humid climatic condition of coastal Odisha because of their high demand during that period. The cultivation of capsicum was tried in winter days of the year 2009–2010. It was observed that the crop yield was more in the greenhouse during off-season as compared to open field condition. The greenhouse with shade net was a suitable protected condition for better plant growth and higher yield compared to without shade net for less variation in temperature due to partial elimination of incoming radiation by the shade net during day hours and prevention of the radiative losses to the cold night sky thus maintaining a better heat distribution inside the greenhouse. The yield of capsicum per square meter of the cultivated area in the greenhouse was 2.34 times more than open field condition. Overall growth of capsicum in terms of height of plants and number of leaves per plant inside the greenhouse was more compared to the open field. Early flowering and fruiting were also observed in the greenhouse condition. The benefit cost ratio for capsicum in the greenhouse was 2.98:1 whereas it was 0.80:1 in case of open field condition. In this naturally ventilated type of greenhouse, the small and marginal farmers of Odisha will be able to grow other vegetables during off-season which would be quite remunerative.

9.1   INTRODUCTION

Cultivation of crops is mainly climate dependent under normal conditions. Hence, all vegetables have their own seasons in which they can be grown. But with the introduction of greenhouse technology [37], farmers can grow vegetables during off-season to fetch a good market value. As there are many small and marginal farmers in Odisha, hence the suitability of a low-tech naturally ventilated greenhouse was evaluated for off-season cultivation of capsicum in coastal Odisha because of its high demand during pre-summer period.

During winter season of 2009, the cultivation of bell pepper was evaluated with three dates of sowing to evaluate and compare its different growth parameters, yield, and yield attributing characters and to harvest during pre-summer period as an off-season vegetable. Looking into the demands of capsicum during off-season and importance of marinating suitable temperature and judicious use of water inside the greenhouse for the growth of capsicum, experiments were conducted under greenhouse and open field conditions.

Production of vegetable crops under protected conditions (inside greenhouse) involves protection of growth stages of vegetables mainly from adverse environmental conditions such as temperature, high rainfall, hail storms, scorching sun etc. Protected conditions for vegetable production are created locally by using different types of structures. These structures are designed based upon climatic condition requirements of the area. Besides temperature, wind velocity, and soil conditions also play major role in the design of inside greenhouse for growing. Therefore, in the present scenario of perpetual demand for vegetables and drastically shrinking land holding, protected cultivation of vegetable crops suitable for domestic as well as commercial purposes is the best alternative for using land and other resources more efficiently.

Green house technology is more suited to vegetable crops (such as tomato, eggplant, cauliflower, capsicum, cabbage, chilly, ladies finger, spinach etc.), flowers (like rose, gerbera, carnation etc.) and nursery for all vegetable crops, because of their small life-span. This technology is mainly suitable for commercial farming, as it requires high initial investment in setting up the entire framework. It involves a structure similar to a house, in which only sunlight is allowed to enter. It consists of a frame or the basic structure made of galvanized iron pipes, bamboo, woods, and iron rods; and the cladding or covering material made of glass and plastic films.

Green house technology is not a new concept to farmers and agri-business entrepreneurs. It has been regularly used in floriculture industry and vegetable seed producing industries. To develop the commercial and hi-tech horticulture in India, Government has been giving support and subsidy to entrepreneurs who are adopting greenhouse technology, through the concerned nodal body, the National Horticulture Board.

Apart from the convenience of growing any crop in any season, green house technology reduces the consumption of irrigation water and fertilizer and infestation of pests and diseases. Considering the reduction of the cultivable land and non-availability of work force at the right time, Indian entrepreneurs can adopt greenhouse technology to grow export oriented fruits and vegetables.

Based on the cost and investment, green house can be categorized in three groups namely low cost green house, medium cost green house, and hi-tech green house.

(a) *Low cost or low-tech greenhouse* is a simple chamber made of polythene sheet of 150–200 microns. It is constructed with locally available materials such as bamboo and timber etc. Unlike conventional or high tech greenhouses, no specific control devices for regulating environmental parameters inside the greenhouse are provided. This type of greenhouse is mainly suitable for cold climatic zone. The temperature within greenhouse increases by 6–10 °C than the outside temperature. It can be adopted for vegetable cultivation during winter season.

(b) *Medium-tech greenhouse* is constructed by using galvanized iron (GI) pipes for permanent framed structure. The greenhouse has single layer covering with UV-stabilized polythene sheet of 200-micron thickness and greenhouse cover is attached to the frame with the help of screws. Whole structure is firmly grouted into the ground to withstand the disturbances against wind. Exhaust fans along with thermostat are provided to control the temperature. Evaporative cooling pads and misting arrangement can maintain a favorable temperature and humidity inside the greenhouse. The greenhouse frame and glazing material have a life span of about 20 years and 3 years, respectively. These types of greenhouses are suitable for dry as well as warm climatic zones and can also be adopted for vegetable production.

(c) *High-tech greenhouse* has a sensor, a comparator, and an operator. The temperature, humidity, and light are automatically controlled. These are indicated through sensor or signal-receiver. Sensor measures the variables, compares the measurement to a standard value, and finally recommends running the corresponding device. This type of greenhouse is suitable for any climatic condition. The horticultural crops, which are sensitive to temperature and humidity, can be grown in high tech greenhouses.

Under prevailing economic conditions, where capital is a scarce input, the choice of majority of the nurserymen and entrepreneurs is in favor of low and medium cost greenhouses, which have partial control over the environmental factors, and are cost effective.

9.1.1   PRINCIPLE OF GREENHOUSE

A greenhouse is generally covered with a transparent material such as polythene or glass. Depending upon the cladding material and its transparency, major fraction of sunlight is absorbed by a vegetable crop and other objects inside the greenhouse. These objects in greenhouse in turn emit long wave thermal radiations for which cladding material has lower transparency. Solar energy is trapped and raises the temperature inside the greenhouse. This is popularly known as greenhouse effect. This rise in temperature in greenhouse is responsible for growing vegetable in cold climates. During summer months, air temperature in greenhouse is brought down by providing cooling devices. In commercial greenhouses besides temperature-controlled humidity, carbon dioxide, photoperiod, soil temperature, plant nutrients etc. facilitate round the year production of a desired vegetable crop. Controlled climatic and soil conditions provide an opportunity to have the maximum potential of a vegetable crop. Greenhouses are framed or inflated structures covered with transparent or translucent material large enough to grow crops under partial or fully controlled environmental conditions to get higher growth and productivity.

9.1.2   BENEFITS OF GREENHOUSE

9.1.2.1   VEGETABLES FOR DOMESTIC CONSUMPTION AND EXPORT

During winter period, the temperature and solar radiations are sub-optimal for growing off-season vegetables namely tomato, capsicum, eggplant, cucumber, okra, and chilly. In tomato, low temperature and low radiation cause puffiness and blotchy ripening. Hence during extreme conditions of winter season (October–February), these vegetables will be cultivated under polyhouse. In a medium cost greenhouse, a yield of tomato and capsicum is about 98.6–110.5 tons/ha and 87.2 tons/ha, respectively. A polyhouse can be made, which will receive sunlight for growing chilly, tomato, eggplant, capsicum, and cucumber. The improved varieties and hybrids of these crops need to be evaluated. The high priced vegetables (asparagus, broccoli, leek, tomato, cucumber, and capsicum) are most important crops for production around metropolis and big cities during winter season or off-season. Thus during winter, it may be useful to grow tomato and capsicum in plastic tunnels as the plants which are protected from cold and frost will manifest faster and better growth resulting in earlier fruiting than the crops grown in the open field.

9.1.2.2   RAISING OFF-SEASON NURSERIES

The cost of hybrid seeds is very high. So, it is necessary that every seed must germinate. For 100% germination, it requires controlled conditions. The cucurbits are warm season crops. They are sown in last week of March to April when night temperature is around 18–20 °C. But in polyhouse, their seedlings can be raised during December and January in polythene bags. By planting these seedlings during end of February and first week of March in the field, their yield can be taken in one and a half months in advance than the normal method of direct sowing. This technology fetches the bonus price due to marketing in the off-season. Similarly, the seedlings of tomato, chilly, capsicum, eggplant, cucumber, cabbage, cauliflower, and broccoli can be grown under plastic cover protecting them against frost, severe cold, and heavy rains. The environmental conditions particularly increase the temperature inside polyhouse and hasten the germination and early growth of warm season vegetable seedlings for raising early crops in spring period. Vegetable nursery raising under protected conditions is becoming popular throughout the country especially in hilly regions. Management of vegetable nursery in protected structure is easier and early nursery can be raised. This practice eliminates danger of destruction of nurseries by hailstorms and heavy rains. Protection against biotic and abiotic stresses also becomes easier.

9.1.2.3   MANIFOLD PRODUCTIVITY IN GREENHOUSES IN COMPARISON TO GROWING THE VEGETABLES IN OPEN FIELD.

9.1.2.3.1   Vegetable Seed Production

Seed production in vegetables is a limiting factor for cultivation of vegetables. The vegetables require specific temperature and other climatic conditions for flowering and fruit setting. Seed production of eggplant, capsicum, cauliflower, and broccoli is very difficult under open conditions. To reduce variation in micro climatic conditions, a protected environment is essential. Therefore, the seed production of highly remunerative crops namely tomato, capsicum, and cucumber is performed under protected environments. The maintenance and purity of different varieties can be achieved by growing these under greenhouse without giving isolation distance particularly in cross-pollinated vegetables namely onion, cauliflower, and cabbage. Hence, vegetable production for domestic consumption and export in low and medium cost greenhouses is feasible for vegetable entrepreneurs in

India. Such production system has not only extended the growing season of vegetables and their availability but has also encouraged conservation of different rare vegetables.

9.1.2.3.2 Hybrid Seed Production

In the twenty-first century, protected vegetable production is likely to be commercial practice not only because of its potential but also out of sheer necessity. In vegetable production, hybrids seeds, transgenic, stress resistant varieties, and synthetic seeds are likely to replace conventional varieties.

9.1.2.3.3 Greenhouse Cultivation of Rare Medicinal, Aromatic, and Ornamental Plant Species

The wide variety of orchids, herbs, rare medicinal and aromatic plants have tremendous scope for export. Greenhouses provide the right type of environmental conditions for conservation, intensive cultivation, and exploitation of these rare species.

9.1.2.3.4 Cultivation in Problematic Regions and Extreme Climates

Even if a fraction of the areas under barren, uncultivable, and fallow conditions can be brought under greenhouses, it would generate substantial income and round the year employment for the farmers in the regions. In cold deserts where no vegetables were grown, nowadays lots of vegetables are being grown due to use of several thousand greenhouses. Greenhouses can also be constructed on lands not suitable for cultivation and on rooftops of residential buildings for growing high value crops.

9.1.2.3.5 Polyhouse for Plant Propagation

Asparagus, sweet potato, pointed gourd, and ivy gourd are sensitive to low temperature. The propagating materials of these vegetables can be well maintained under polyhouse in winter season before planting their cuttings in early spring-summer season for higher profit.

9.1.3   GREENHOUSES: WORLD SCENARIO

There are more than 50 countries now in the world where cultivation of crops is undertaken on a commercial scale under protected cover. USA has a total area of about 4000 ha under greenhouses mostly used for floriculture with a turnover of more than 2.8 billion US $ per annum and the area under greenhouses is expected to go up considerably, if the cost of transportation of vegetables from neighboring countries continues to rise. The area under greenhouses in Spain has been estimated at 25,000 ha and Italy 18,500 ha used mostly for growing vegetable crops like watermelon, capsicum, strawberries, beans, cucumbers, and tomatoes. In Spain, simple tunnel type greenhouses are generally used without any elaborate environmental control equipments mostly using UV-stabilized polyethylene film as cladding material. In Canada, the greenhouse industry caters both to the flower and off-season vegetable markets. The main vegetable crops grown in Canadian greenhouses are tomato, cucumbers, and capsicum. Hydroponically grown greenhouse vegetables in Canada find greater preference with the consumers and could be priced as much as twice the regular greenhouse produce.

The Netherlands is the traditional exporter of greenhouse-grown flowers and vegetables all over the world. With about 89,600 ha under cover, the Dutch greenhouse industry is probably the most advanced in the world. Dutch greenhouse industry however relies heavily on glass-framed greenhouses, in order to cope up with very cloudy conditions prevalent all the year round. A very strong research and development component has kept the Dutch industry in the forefront. The development of greenhouses in Gulf countries is primarily due to the extremity in the prevailing climatic conditions. Israel is the largest exporter of cut flowers and has wide range of crops under greenhouses (15,000 ha) and Turkey has an area of 10,000 ha under cover for cultivation of cut flowers and vegetables. In Saudi Arabia, cucumbers and tomatoes are the most important crops contributing more than 94% of the total production. The most common cooling method employed in these areas is evaporative cooling. Egypt has about 1000 ha greenhouses consisting mainly of plastic covered tunnel type structures. Arrangements for natural ventilation are made for regulation of temperature and humidity conditions. The main crops grown in these greenhouses are tomatoes, cucumbers, peppers, melons, and nursery plant material.

Asia, China, and Japan are the largest users of greenhouses. The development of greenhouse technology in China has been faster than in any other country in Asia. With a modest beginning in late seventies, the area under greenhouses in China has increased to 48,000 ha in recent years. Out of this 11,000 ha is under fruits like grapes, cherry, Japanese persimmon, figs, loquat, lemon, and mango. The majority of greenhouses use local materials for the frame and flexible plastic films for glazing. Most of the greenhouses in China are reported to be unheated and use straw mats to improve the heat retention characteristics. Japan has more than 40,000 ha under greenhouse cultivation of which nearly 7500 ha is devoted to only fruit orchards. Greenhouses in Japan are used to grow wide range of vegetables and flowers with a considerable share of vegetable demand being met from greenhouse production. Even a country like South Korea has more than 21,000 ha under greenhouses for production of flowers and fruits. Thus, greenhouses permit crop production in areas where winters are severe and extremely cold as in Canada and USSR, and also permit production even in areas where summers are extremely intolerable as in Israel and Kuwait. Greenhouses in Philippines make it possible to grow crops in spite of excessive rains and also in moderate climates of several other countries. Thus, in essence, greenhouse cultivation is being practiced and possible in all types of climatic conditions. The approximate area under greenhouse cultivation in different countries is given in Table 9.1.

TABLE 9.1   Approximate Area (ha) Under Greenhouses for Different Countries [28].

Country	Area (ha)	Country	Area (ha)
Japan	54,000	Turkey	10,000
China	48,000	Holland	9600
Spain	25,000	USA	4000
South Korea	21,000	Israel	1500
Italy	18,500	India	1000

India has very little area under greenhouses as is evident from Table 9.1. The major share has been in the Leh and Ladakh regions of Jammu and Kashmir where commercial cultivation of vegetables is being promoted.

In Northeastern hilly region, polyhouse cultivation is still a new emerging technology for raising nursery of vegetable crops. Assistance provided under the plasticulture scheme since the VIII & IX plan has helped in generating awareness about the importance of greenhouses in enhancing productivity and production, particularly of horticultural crops. Out of 1000 ha area under greenhouses in India, 183 ha has been covered in the Northeastern states, the maximum area being in Sikkim.

9.1.4   STATUS OF GREENHOUSES IN INDIA

While greenhouses have existed for more than one and a half centuries in various parts of the world, in India, the use of greenhouse technology started only during 1980s and it was mainly for research activities. This may be because of our emphasis, so far had been on achieving self-sufficiency in food grain production. However, in recent years in view of the globalization of international market and tremendous boost that is being given for export of agricultural produce, there has been a spurt in the demand for greenhouse technology. The National Committee on the use of Plastics in Agriculture (NCPA) has recommended location specific trials of greenhouse technology for adoption in various regions of the country.

Greenhouses are being built in the Ladakh region for extending the growing season of vegetables from 3 to 8 months. In the northeast, greenhouses are being constructed essentially, as rain shelters to permit off-season vegetable production. In the Northern plains, seedlings of vegetables and flowers are being raised in the greenhouses either for capturing the early markets or to improve the quality of the seedlings. Propagation of difficult-to-root free species has also been found to be very encouraging. Several commercial floriculture ventures are coming up in Maharashtra, Tamil Nadu, and Karnataka states to meet the demands of both domestic and export markets. The commercial utilization of greenhouses started from 1988 onwards and now with the introduction of Government's liberalization policies and developmental initiatives, several corporate houses have entered to set up 100% export oriented units. In just four years, since implementation of the new policies in 1991, 103 projects with foreign investment of more than Rs. 800 million have been approved to be set up in the country at an estimated cost of more than Rs.10,000 million around Pune, Bangalore, Hyderabad, and Delhi. Thus the area under climatically controlled greenhouses of these projects is estimated to be around 300 ha. Out of which many have already commenced exports and have received very encouraging results in terms of the acceptance of the quality in major markets abroad and the price obtained.

9.1.5   CONSTITUENTS OF CONTROLLED ENVIRONMENT

9.1.5.1   TEMPERATURE

Optimum temperature refers to the most suitable temperature at which plant can grow under particular type of climatic condition. Optimum temperatures for different crops are different [24]. Each crop has an optimum temperature at which enzymes, which are heat sensitive and responsible for bio-chemical reactions, are most active.

Net growth of crop occurs when photosynthesis exceeds respiration. Hence, in order to achieve high levels of photosynthesis rate, plant temperatures are kept low at night to decrease respiration rate and warmer during day to increase photosynthesis [33, 37].

Ali and Abdulla [1] studied the environmental conditions for tomato and cucumber in a greenhouse for hot and dry arid lands. They observed that the yield of tomato was two times higher than outside conditions when greenhouse temperature was maintained at the temperature range of 16–30 °C throughout the year whereas outside air temperature varied from 8.5–45.5 °C.

9.1.5.2   LIGHT INTENSITY

The intensity of incoming solar radiation is an important parameter for influencing the photosynthetic activity of plants. The light intensity varies from place to place but it generally varies from zero at the beginning of the day to about 90,000–100,000 lux (lumen/m²) around noontime. Light intensity on cloudy days is quite low which leads to poor photosynthetic process. Light intensity below 3200 lux and above 129,000 lux is not ideal for plant [37]. Hence solar radiation transmittance needs utmost attention while designing and constructing the greenhouse. It is also influenced by the orientation of the greenhouse and the sun elevation. Greenhouses with curved roofs have better transmittance than greenhouses with a pitched roof of 25° slope.

During peak summer, some protection from high intensity of light may be required because the high intensity raises the temperature of leaf and causes sun burning. Therefore, some type of shading screen, either over the greenhouse or inside greenhouse is provided. Spraying the greenhouse cover with a suitable shading compound such as limewater, white latex, and paint with water is recommended during summer [37].

9.1.5.3   HUMIDITY

Like light and temperature, humidity is also an important parameter in the greenhouse climate. Absolute humidity is the amount of water vapor actually present in the air. Relative humidity inside the greenhouse should be between 60–70% for better growth of plants [6]. If the plants have a well-developed root system, then relative humidity above 40% is preferred to avoid water stress conditions. Even very low relative humidity (less than 20%) can cause wilting due to higher rate of evaporation from plant. High levels of humidity can lead to yield loss for tomato crop [7, 25]. Higher humidity (above 80%) also leads to occurrence of fungal diseases within the greenhouse. Jolliet [29] had optimized the humidity and transpiration in the greenhouse and he reported that 60–70% relative humidity was congenial for the plant for its desirable growth in the greenhouse.

9.1.5.4   CARBON DIOXIDE

Carbon dioxide is an important parameter for plant growth like water, light, soil nutrients, and temperature. In photosynthesis process, a plant leaf seeks to combine molecules of CO_2 and water in the presence of sunlight to form carbohydrates and oxygen as shown in following equation.

$$6CO_2 + 6H_2O \xrightarrow[\text{Photosynthesis}]{\text{Sun light}} C_6H_{12}O_6 + 6O_2 \uparrow \qquad (9.1)$$

Several researchers have indicated that closed greenhouse system offers good opportunity to improve production through the elevation of CO_2 levels. Carbon dioxide, which comprises about 0.03% (300 ppm) of ambient air, is essential for plant growth. This level of carbon dioxide in atmospheric air is sufficient to meet the photosynthetic requirement of open field crops. In the closed field conditions, that is, in greenhouse, the level of carbon dioxide rises up to nearly 1000 ppm, because respired carbon dioxide remains trapped overnight. As the sunlight becomes available, photosynthesis process begins and carbon dioxide from greenhouse air gets depleted. Owing to this, the carbon dioxide level in greenhouse even goes below 300 ppm before noon. If greenhouse air does not receive additional carbon dioxide from any other source, the plant would become carbon dioxide deficient resulting in poor growth. Carbon dioxide enrichment is therefore essential when greenhouse is sealed against infiltration particularly during winter period. Critten [13] reported that crop yield was increased by 20–30% when carbon dioxide level was maintained from 1000–1500 ppm inside the greenhouse The most common method of carbon dioxide supplementation is through burning of carbon fuels. Care should be taken to assure complete combustion by providing outdoor air infiltration to supply adequate oxygen levels for combustion.

9.1.5.5   ROOT MEDIUM

In addition to the above, root medium also plays an important role for cultivation of crops in greenhouse as well in the field. This is also known as growing medium for greenhouse crop cultivation. It must serve as reservoir for plant nutrients. Also it must hold water in a way that is available to plant and at the same time it provides the path for the exchange of gases between roots and the atmosphere above the root medium. Finally the root medium must provide an anchorage or support for the plant. The desirable properties of a root medium are as follows:

a. Stability of organic matter.
b. Maintenance of carbon: nitrogen ratio.
c. Keeping desirable bulk density.
d. Capacity for moisture retention and aeration.
e. Balance of pH level.
f. Higher level of cation exchange capacity.

The growing media of soil with manure and sand with manure are used for raising crop both in pot and in field under greenhouse. Organic matter is mixed in the range of 0–100% with the soil and sand for preparing good root medium.

9.1.6   GREENHOUSE CLIMATE REQUIREMENTS [47]

a. Plants grown under protected cultivation are mainly adapted to average temperatures ranging from 17 to 27 °C. Taking into account the warming-up effect of solar radiation in the greenhouse, this temperature range is possible without any heating arrangement in it when outside ambient temperature prevails in the range from 12 to 22 °C.
b. If the mean daily outside temperature is below 12 °C, greenhouse needs to be heated, particularly at night. When mean daily temperature is above 22 °C especially during summer, artificial cooling is necessary or cultivation in greenhouse is to be stopped. Natural ventilation is sufficient when ambient mean temperatures range from 12 to 22 °C.
c. The absolute maximum temperature for plants should not be higher than 35–40 °C.
d. The minimum threshold for soil temperature is 15 °C.

e.   Verlodt [48] suggests a threshold of the average night temperature as 15–18.5 °C for heat requiring plants such as tomato, pepper, cucumber, melon, and beans.

f.   The safe ranges of relative humidity are from 70–80%.

9.1.7   JUSTIFICATION FOR THIS STUDY

Growth of population and less availability of required food materials like vegetables and fruits (the important sources of vitamins, minerals, fiber etc.) have become the global concerns. In most of the developing countries, traditional open field cultivation has not been able to maintain the sustainable vegetable production. Because open field agricultural practices only control the nature of the root medium through tillage operation, fertilizer application, and irrigation. They do not ensure the control on the environmental parameters such as sunlight, air composition, and temperature that regulate the growth of plant. Hence a large number of winter vegetables, flowers, and other horticultural crops cannot be grown locally during summer and have to be transported from the long distance places as per the needs of the consumers. The same practice also happens for summer crops during winter period. To meet the increased demand of off-season vegetables, greenhouse technology appears to be a promising alternative.

The demand of fresh as well as good quality vegetables and cut flowers at global level is also increasing. This calls for increasing productivity at a higher rate. The increased demand cannot be met through the traditional method of agricultural production. It necessitates improved and new alternative technologies to enhance production under normal as well as adverse climatic conditions and to bridge the gap between demand as well as existing production of vegetables, fruits, and flowers.

Greenhouse, in this regard, helps to create favorable conditions where production of vegetables and flowers is made possible throughout the year or part of the year as per the requirement. It not only creates suitable environment for the plants but also encourages proper growth and fruiting as compared to open field cultivation. The greenhouse technology has also tremendous scope especially for production of hybrid seeds, nursery raising, ornamental plants, medicinal plants which fetch more prices in markets.

In coastal Odisha, the mean air temperature varies from 25 to 37.17 °C in summer, 24.53–32.72 °C in rainy and 14.88–28.33 °C in winter seasons. During the cold season, there is a high demand of capsicum (*Capsicum annuum L.*) vegetable. Capsicum, a rich source of vitamin C is gaining popularity in big cities as a nutritious and export oriented vegetable. But the cultivation of this vegetable is not so easy for want of suitable ambient conditions to promote its growth and development. In Odisha, capsicum is generally grown in open field condition. Though capsicum is a crop of winter season but very low temperature due to cold wind and frost in winter period and very high temperature during summer season retards its growth and yield. Moderate temperature in the range of 22–27 °C [3] during first part of September to end of November is optimum for its cultivation, but the above situation rarely prevails during winter period for the cultivation of capsicum. Hence the controlled environment through the solar greenhouse is the right alternative for the higher production and productivity of capsicum by maintaining the required environmental conditions in winter season. So there is a need to increase the temperature for safe growing of capsicum vegetable as off-season crop in winter season as well as decrease of temperature in the pre-summer periods.

As greenhouse allows faster temperature increase during sunny day and slower temperature decrease in night hours, it is considered to be the most suitable structure for off-season cultivation of these vegetables. But higher operating cost of high-tech controlled greenhouse will be a constraint for popularization of this technique in a state like Odisha where 75–85% of farming community is small and marginal [4]. Hence, there is a need to study the suitability of low-tech naturally ventilated greenhouse in coastal Odisha for off-season cultivation of capsicum.

The control of various environmental parameters inside the greenhouse, suitable for favorable growth of plant can be studied mathematically by developing a suitable thermal model, which is required to optimize those parameters involved in either heating or cooling of greenhouse. The modeling can also be used to optimize greenhouse air temperature (one of the important constituents of the environment inside the greenhouse) for higher yield of a plant inside greenhouse for a given climatic condition. Thermal modeling requires basic energy balance equations for different components of greenhouse system for a given climatic (solar radiation, ambient air temperature, relative humidity, wind velocity etc.) and design (volume, shape, height, orientation, latitude etc.) parameters. Basic knowledge of heat and mass transfer is also of great importance in deriving energy balance equation for heating and cooling operations of a greenhouse under given climatic conditions. The transfer of heat energy occurs as a result of driving force called temperature difference and mass transfer takes place in the form of evaporative heat transfer.

To facilitate the modeling procedure, a greenhouse is considered to be composed of a number of separate but interactive components. These are greenhouse cover, the floor, the growing medium, enclosed air, and the plant. The crop productivity depends on the proper environment and more specifically on the thermal performance of the system. The thermal performance of a greenhouse can be studied with the help of a mathematical model with suitable assumptions. Energy balance equations are derived to formulate the model, which permits the prediction of environmental conditions in a greenhouse from outside atmospheric conditions.

In this study, an attempt has been made to develop a mathematical model based on energy balance equations for each component of the greenhouse. The mathematical model was then validated with the experimental findings for its applicability in enhancing production and productivity of an off season *capsicum* in a given climatic condition with the following objectives,

a. To compare the growth and yield of *capsicum* inside greenhouse and in open field condition.
b. To compare the economics of *capsicum* cultivation inside and outside the greenhouse.

9.2   REVIEW OF LITERATURE

9.2.1   VEGETABLE CULTIVATION IN GREENHOUSE

Scientists all over the world have constantly been doing research for making the greenhouse technology more feasible and cost effective. The research findings in the past are at present helping the other researchers to do more studies in that direction. The work done by the several scientists for vegetable cultivation, greenhouse heating, and thermal modeling has been reviewed.

The evaluation of different vegetable crops grown inside greenhouse was conducted at Central Institute of Agricultural Engineering, Bhopal. It was found that okra and capsicum gave about three times higher yield inside the greenhouse as compared to open field condition. The fruit yield of pepper in greenhouse was two and half times higher under greenhouse. The biomass yield was also higher by five to six times [38].

The raising of seedlings was studied under plastic tunnels of size 4 m × 1.7 m × 0.85 m made of UV-stabilized film of 200 microns. The experiment was conducted during June–August, 1989. Trials on raising of vegetable and rice seedlings suggested a considerable reduction in the time required for the seedling preparations [36].

An attempt was made to grow cucurbits in pots in a controlled environment inside a solar greenhouse using several cooling concepts. An average of 9.22 cucumbers per pot (total average weight 1789 g) was produced during winter season at the experimental site of IIT, New Delhi. The correlation was proposed to predict the yield as a function of number, length, and the perimeter of cucumbers [43].

The cultivation of tomato and eggplant was taken under a semi-cylindrical greenhouse of size 10 m × 4 m during January–June, 2001 at Institutional Farm of College of Technology and Engineering, Udaipur. It was observed that the increase in plat height of tomato and eggplant were, respectively, 30 and 25% after ninth week of transplantation. The increase in dry matter weight of tomato and eggplant was 80 and 88%, respectively, after ninth week of transplantation. The total yield of tomato and eggplant inside the greenhouse were 52 and 42 kg as compared to 14.5 and 24.5 kg at open field, respectively. The increase in the yield inside the greenhouse as compared to open field was 57% for tomato and 71% for eggplant [31].

The use of solar energy [12, 14] for growing capsicum in the pots and in the ground has been studied under controlled environment in a solar greenhouse (IIT model) and in open field during August 2000 to March 2001 [35]. From the experiment, it was found that capsicum is very sensitive to temperature and its growth as well as yield was reduced during off-season that is, in peak winter. Maintaining suitable temperature inside the greenhouse increases the production of capsicum compared to open field condition. Germination of seeds took place nearly one week earlier in the greenhouse than the ambient conditions, because of higher room temperature. The yield of capsicum became almost double in the greenhouse than the outside crop. Heat loss was reduced in nighttime during winter months due to the presence of north wall built of bricks and by the use of movable insulation.

Performance of tomato varieties were evaluated under organic farming in greenhouse as well as in field conditions during the winter season in Tamil Nadu. Observations regarding plant height, fruit number, average fruit weight, and yield were recorded for the varieties Pusa Ruby, Rama, and PKM-1. It was found that *Pusa Ruby* was suitable for greenhouse condition with a yield of 666 g/plant followed by Rama with the yield of 413 g/plant [15].

Studies on growth and yield of cucumber and broccoli were undertaken under a 20 m × 5 m steel pipe framed greenhouse glazed with 200-micron UV-stabilized film at Division of Vegetable Crops, IARI, New Delhi. The greenhouse was fitted with an evaporative cooling system consisting of two exhaust fans of 45 cm size at one end and 5 m × 1.3 m cooling pad at the other end. It was concluded that the maximum yield of 125.82 t/ha was recorded at 60 cm × 60 cm spacing of cucumber with pruning of all primary branches after two nodes. Similarly the maximum yield of 21.06 t/ha was achieved by planting broccoli seedling at spacing of 30 cm × 50 cm under greenhouse condition [30].

The growth of winter-grown tomatoes was evaluated under low tunnels of 200, 300, and 400 micron polyethylene film at G. B. Pant University of Agriculture and Technology, Pantnagar, Uttarakhand. The increase in minimum and maximum temperature inside tunnels as compared to open was generally in the range of 0.5–1.5 °C and 5–7 °C, respectively, during November to January month. In general, all types of polyethylene covering produced significantly higher plant height and number of primary branches as compared to uncovered treatment. The shading might have increased the cell elongation which increased the length of internodes resulting in increased plant height. The improved vegetative growth under the low plastic tunnels largely attributed the increase in air temperature and relative humidity inside the tunnels. In addition to increase in air temperature, it was likely that increase in soil temperature and moisture were factors favorable to growth under polyethylene covers [32].

A study was undertaken to develop some techniques for quick emergence of paddy seedlings in winter. Four techniques were tried in the experiment: transparent plastic low tunnel, white plastic strip over the crop row, and black plastic strip over the crop row and paddy straw cover. Of which, transparent plastic low tunnel technique was found to be appropriate as the seedlings could be raised within 25–28 days against 40–45 days taken in the open field thereby saving of about 15–20 days. Temperature inside the tunnel was found about 4 °C more than the control. The chlorophyll content of the seedlings was also 35% more compared to open field seedlings [40].

A study was conducted in a semi-cylindrical fan-pad greenhouse of 100 m² size on the performance evaluation of rose revealed that both better quality and quantity of flowers could be obtained from the greenhouse compared to the open field cultivation under composite climate of Navsari, Gujarat. The planting geometry of 0.5 m × 0.33 m gave very good results in terms of production of maximum number of flowers per unit area (152.82 flowers/m²) and maximum stalk length of 53 cm. The study revealed the increase in rose production by 52.05% and the net income by 838.91% over open field cultivation in a normal market condition [39].

A study was carried for creating suitable environment for the germination and subsequent growth of plant in the greenhouse of size 7 m × 3 m × 2 m for raising early summer vegetable nursery at PAU, Ludhiana. It was observed that the average air temperature inside the greenhouse was 10–12 °C higher than the ambient air temperature. Inside, the average soil temperature was also 5–7 °C more than the corresponding temperature outside the greenhouse. Greenhouse microclimate was modified by covering its roof with a polyester sheet to cut down the effect of night sky radiation thereby raising the inside minimum temperature. It was observed that the germination of muskmelon seeds sown inside the greenhouse occurred one week earlier as compared to the sown in the open field. The seedlings inside the greenhouse took only three weeks to attain two leaf stages whereas seedlings sown in the open field took five weeks to reach up to two-leaf stage. Thus, there was a clear saving of 15 days in raising the nursery under the greenhouse [42].

9.2.2   HEATING OF GREENHOUSE

Heating of greenhouse is one of the most important activities during winter season. The passive heating may be due to water storage, rock bed storage, presence of north wall, mulching, phase changing materials, and movable insulation. In active heating, warm water, ground air collector, earth to air heat exchanger, and underground geothermal water may be a better option for thermal improvement.

Movable insulations are usually night curtains or thermal screens, which are drawn inside or outside the greenhouse cover during nighttime in winter months to reduce heat losses to ambient resulting in the conservation of energy in the greenhouse. These movable insulations are uncovered during daytime in order to allow solar radiation into the greenhouse for thermal heating. Chandra and Albright [11] have analytically determined the effects of night curtain on the heating requirement of greenhouse and predicted that nearly 70% of heating load could be saved by use of night curtain. Garzoli and Blackwell [16] studied the effects of movable insulation that could check the exchange of long infrared radiation, emitted by the roofing material with sky during cold night. Barral et al. [9] tested the performance of integrated thermal improvements of thermal curtains as well as thermal blankets and reported that these movable insulations were proved to be very efficient to provide the required temperature levels for the healthy growth of tomatoes and peppers during winter period. Plaza et al. [41] also reported energy saving of 20% in the greenhouse by the use of thermal insulation.

Grafiadellis and Kyritsis [21] studied the impacts of two solar heating systems for thermal heating of greenhouse in Greece. The first system consisted of three major elements: (a) the external solar energy collector (b) an underground water reservoir, and (c) plastic tubes of black polyethylene through which warm water was circulated to heat the greenhouse. On cold nights, the temperatures of air and soil in the solar heated greenhouse were, respectively, 6–5 °C and 8–10 °C higher than those of non-heated greenhouse. The second system consisted of an under-ground water reservoir, a fan, and a pump. The solar heating systems were found to be very useful for plastic greenhouses in Greece, since they protected the crop from the low temperature conditions in the non-heated greenhouses. Also in the solar heated greenhouse, the production of tomato was increased by 30–50% and the harvesting of tomatoes was done 2–3 weeks earlier than that of the non-heated greenhouse.

Arinze et al. [5] have studied the effects of movable internal curtain, placed in between the glazings of a double-layered greenhouse. They developed a computer program to predict the performance of the thermal curtain in reducing the energy consumption of a greenhouse during winter period. The comparisons between the experimental and computer results showed a high degree of correlation. Their results indicated that under average climatic condition, the heating requirements of a double-glazed greenhouse using thermal curtain could be reduced as much as 60–80%.

Grafiadellis [20] investigated the use of underground water in a polyethylene-covered greenhouse of 150 m^2 area. The warm water was sprayed at the rate of 5 m^3/h at night from the nozzles located on the ridge of the greenhouse. At the same time, the air inside the greenhouse was circulated by a blower in order to absorb heat from the sprayed water as well as from the floor, which was heated by solar radiation during daytime. The system was proved to be simple and effective in raising the temperatures of air in the greenhouse to the levels of 8 °C higher than those of an unheated conventional greenhouse.

Al-Amri and Ali [2] studied the effect of solar water heater. It was fixed on the interior of the gable of an even span greenhouse. He reported that the productivity of tomato was enhanced by 46.67% in the greenhouse due to heating of the greenhouse by solar water heater.

The use of shallow solar pond in greenhouse for cold climatic conditions was studied by Hussaini and Suen [26] for heating of greenhouse. A shallow solar pond built as an integral part of a greenhouse, can be used for both collection and storage of solar energy. The heat given out by the shallow solar pond is used for heating the greenhouse during the day. In addition, the collected heat during the day can also be used efficiently for heating the greenhouse at night. The performance of the shallow solar pond was investigated under different weather conditions for different solar pond areas. The results showed that the use of a shallow solar pond in a greenhouse could provide great savings of energy needed to heat the greenhouses during winter months.

Barral et al. [9] have studied the performance of a greenhouse of area 105 m², integrated with thermal curtains in walls and roof, thermal blanket over the plant mass and circulation of underground geothermal warm water through the polyethylene tube positioned on its floor. They reported the system to be very efficient to maintain the temperature of air above 13 °C inside the greenhouse for the healthy growth of tomatoes and green peppers during winter period.

Plaza et al. [41] have also reported 20% energy saving in the greenhouse by the use of thermal insulation.

Bargach et al. [8] conducted experiment by using solar flat-plate collectors for heating of a tunnel greenhouse. Heating system consisted of coldwater tank, flat-plate collector, hot-water tank and the heat exchanger tubes installed inside the greenhouse. A model was developed to simulate the thermal behaviors of the system. The maximum temperature of water in the inlet and outlet ends of the solar collector was found to be 20 and 40 °C, respectively, during the month of January. The mean thermal efficiency of the system was 55.5%.

Singh and Tiwari [46] have developed a thermal model to study the effects of storage north wall integrated with the ground air collector on the variations of plant and greenhouse air temperature. From the results, it was observed that the temperature of air inside the greenhouse was nearly 8 °C higher than the ambient air temperature during nighttime.

Kurpaska and Slipek [34] studied the efficiency of two substratum-heating systems in the greenhouse. In horticultural practice, two substrate-heating systems predominate. One is the buried pipe heating system in which heating tubes are placed inside the substrate with circulating warm water and the other is the vegetation heating system where heating tubes are placed directly upon the soil surface. They evaluated the optimum values of the decision variables such as the depth and span of the heating elements in the soil substratum and also the temperature of the medium fed into the heating system. The basic analysis showed that for similar substrate temperature conditions, the vegetation heating system required 3 °K higher water temperature than the buried pipe system and the heat loss was higher for the vegetation system.

A mathematical model along with its experimental validation was formulated by Jain and Tiwari [27] to study the thermal behavior of a greenhouse combined with a ground air collector for the heating of greenhouse. The results of the experiments conducted during December 2000 to March 2001 in the climate of Delhi, India for a greenhouse (24 m² area) with a brick north wall and ground air collector arrangement showed that the minimum temperature of air during nighttime was maintained above 14 °C in the greenhouse for creating congenial environment for healthy growth of the crop.

9.2.3   THERMAL MODELING

Different models have been developed to study the effects of different controlling parameters for enhancing the effectiveness of a greenhouse. Models are generally used as a tool to study the behavior of those controlling factors for a greenhouse in different climatic conditions and its various designs.

Attempt was made to apply an analytical expression to the plants and enclosed air in a greenhouse for various design parameters and a given climatic condition. A numerical method was used to validate the analytical expression for the plant temperature. The analysis was based on energy-balance equations for different components of the greenhouse. Numerical computation was carried out for a typical summer day in Delhi. The effects of parameters such as rate and the duration of ventilation, movable insulation etc. were studied. Finally the model was used to standardize a greenhouse for any climatic conditions [44].

A thermal model was developed to study the effects of storage north wall integrated with ground air collector on the variation of plant and green house air temperature. From the results, it was observed that the temperature of the air inside the green house was nearly 8 °C higher than the ambient air temperature during nighttime [46].

A mathematical model was developed to study the effect of various energy-conservation measures to arrive at a set of design features for an energy efficient greenhouse. The simulation results indicated that under cold climatic conditions of northern India, a gothic arch shaped greenhouse required 2.6 and 4.2% less heating as compared to gable and Quonset shapes, respectively. An east–west oriented Gothic arch greenhouse required 2% less heating as compared to north-south oriented one. The use of night curtains reduced the nighttime heating requirement by 70.8% and daily requirement by 60.6%. By replacing the single cover on the southern side with inflated double wall glazing, the heating requirement was reduced by 23%. The combination of the design features for an energy efficient greenhouse suitable for cold climatic conditions was found to reduce the greenhouse heating needs by 80%. An internal rock bed thermal storage system met the remaining heating energy requirements of the energy-conserving greenhouse [22].

A computer model was developed on transient analysis of the greenhouse to predict the room air temperature, storage water temperature and the thermal energy storage effect of water mass in a low cost passive greenhouse. Analytical expression based on an energy balance for each component was in terms of climatic as well as design parameters. It was observed that (a) there was a significant thermal energy storage effect of the water mass on room temperature and (b) thermal load leveling which was found to decrease with an increase in the mass of storage water varied with month of year. The predicted room and water temperature showed fair agreement with experimental values [23].

A study was carried on characterization and modeling of the most relevant convective transfers contributing to the description of the distributed greenhouse climate. It focused on the study of the distributed climate which required the equation governing the fluid flow. A complete study pertaining to air movement inside the greenhouse was presented and it was particularly focused on studies dealing with plant air interactions particularly the leaf boundary layer climate and the air flows within the canopy. The computer fluid dynamic (CFD) software became more realistic and was able to describe the main features of the distributed climate inside the greenhouse with fair accuracy [10].

The design quality of a sunlight greenhouse was evaluated in terms of its practicality, economics, creativeness, and artistry. The factors affecting greenhouse design quality were analyzed that are, function, budget cost, traditional culture, and others. A mathematical model for comprehensive evaluation by fuzzy theory was established carrying the evaluation for a big span sunlight greenhouse. The practice proved that this method was simple, suitable, and reliable [45].

A mathematical model was developed to study the thermal behavior of a greenhouse while heating with a ground air collector. A computer program based on MATLAB software was used to predict the plant and room temperature as a function of various design parameter of the ground air collector. Experiments were conducted during December 2000 to March 2001 for an even span greenhouse of effective floor area of 24 m^2 with ground air collector and having a brick north wall. The model was validated experimentally in the climate of Delhi for the winter season. A parametric study involved the area of the ground air collector, mass flow rate, and heat capacity. The predicted plant and room temperature showed fair agreement with the experimental values [27].

9.3   MATERIALS AND METHODS

Looking into the demands of capsicum during off-season and importance of marinating suitable temperature inside the greenhouse for the growth of capsicum, experiments were conducted under greenhouse and open field conditions. The germination period and different growth parameters such as height of plant, weight of biomass and yield of the plants both inside and outside the greenhouse were observed. To validate the thermal model developed for the environmental conditions in the greenhouse, air temperature, plant temperature and solar radiations were also measured. The materials used and the methods followed to study the viability of off-season cultivation of capsicum under naturally ventilated greenhouses in coastal region of Odisha are presented in this section.

9.3.1   EXPERIMENTAL SITE

The experiment was conducted inside the nursery site of the Department of Horticulture, Orissa University of Agriculture and Technology, Bhubaneswar during 2009–2010. The location is situated at 20°15`N latitude and 85°52`E longitude with an elevation of 25.9 m above the mean sea level and nearly 64 km west of the Bay of Bengal.

9.3.2   DETAILS OF THE GREENHOUSE AND GREENHOUSE WITH SHADE NET

A semi-circular shaped greenhouse (Fig. 9.1) covering floor space of 4 m × 12 m (48 sq m) oriented in east-west direction was used for this study. The greenhouse was covered with ultra violet (UV) low-density polyethylene (LDPE) film of 200-micron thickness. The greenhouse was covered with a Netlon-made shade net of 50% as a shading device as when required.

9.3.3   GREENHOUSE VENTILATION

To ventilate the greenhouse with cool air to bring down greenhouse environment (air temperature and humidity) to the desired level, both sides of the polyethylene sheet were rolled upward to control the partial thermal environment.

9.3.4   MEASUREMENT OF GREENHOUSE PARAMETERS

The following parameters were measured inside and outside the greenhouse:

a. Solar radiation at selected points on the wall and on the floor of the greenhouse.
b. Total and diffuse radiation on the horizontal surface outside the greenhouse.
c. Greenhouse enclosure (greenhouse air) temperature and the outside ambient temperature.
d. Plant temperature inside and outside the greenhouse.
e. Relative humidity inside the greenhouse.
f. Growth of the plant and fruit.
g. Total yield of the crop.

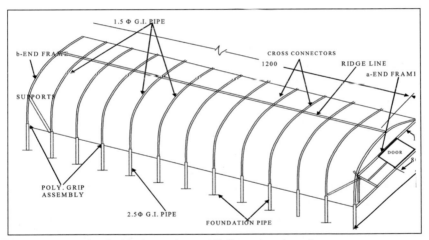

FIGURE 9.1   Semi cylindrical greenhouse (all dimensions in cm).

9.3.5   INSTRUMENTS FOR MEASUREMENT OF PARAMETERS

9.3.5.1   LUXMETER

The luxmeter weighing 160 g and of LUTRON make having a display of 13 mm LCD was used for measuring solar light intensity in lux within a range of 0–50,000 in three ranges. It generally measures the total solar radiation in W/m². However, the diffuse solar radiation was also measured manually by providing a shade over its photovoltaic sensor.

9.3.5.2   TEMPERATURE MEASURING EQUIPMENT

Thermometer measures the temperature of a substance. Number of techniques are used in thermometers depending on factors such as the degree of accuracy required and the range of temperature to be measured, but they all measure temperature by making use of some properties of a substance that vary with temperature. Air temperature was measured with mercury thermometer. The least count of the thermometer used was 1.0 °C. The thermometer was placed in the middle of greenhouse at a height of 1.5 m from the ground. It was placed half way from the floor to roof and its bulb was shaded from direct sunlight. The other thermometer to measure the ambient temperature was placed outside at approximately the same height from the ground and about 4 m away from the greenhouse in a shaded place.

9.3.5.3   PLANT TEMPERATURE

The plant or leaf temperature was measured with the help of an Infrared thermometer. This was of digital type and had a least count of 0.1 °C. The minimum distance between the plant leaf and instrument was kept as 1 m. It directly displayed the readings for plant temperature, when the trigger was pressed.

9.3.5.4   MFASUREMENT OF RELATIVE HUMIDITY

A portable dial type hair hygrometer (Huger make, Germany) was used for measuring the relative humidity (γ) inside the greenhouse. The hygrometer had a least count of 1%. It was placed at the center of the greenhouse at 1.5 m above the ground. This hygrometer works on the principle that the length of hair inside the hygrometer changes with the change of relative humidity by absorption of the moisture. The hair is connected to the pointer through a coil, which moves on the dial showing direct reading of relative humidity.

9.3.5.5   MEASUREMENT OF GROWTH OF PLANT

The growth of plant, both, in the greenhouse as well as in the field was measured with the help of a measuring tape and scale. The height of the sample plants of each treatment plot was measured at 15, 30, 45, and 60 days after sowing and at final harvest with the help of meter scale and expressed in centimeter (cm).

9.3.5.6   MEASUREMENT OF WEIGHT

The digital electronic balance (Sansui Electronics Pvt. Ltd., New Delhi, India) was used to measure the weight of fertilizer/chemicals and the fruits of capsicum. The weighing capacity of this balance was 1 kg with least count of 0.1 g. It was operated by 230 volt, AC supply.

9.3.6   EXPERIMENTAL DESIGN

The experiment was laid out in a Complete Randomized Design (CRD) with six different treatment combinations of factor A and factor B in four replications. The details of layout are shown below.

Layout for capsicum plantings				
Greenhouse (G1)		Sowing date	Open field (G2)	
R_1	R_3		R_1	R_3
R_2	R_4	S_1	R_2	R_4
R_4	R_1		R_4	R_1
R_2	R_3	S_2	R_2	R_3
R_3	R_4		R_3	R_4
R_1	R_2	S_3	R_1	R_2

Size of the greenhouse: 12 m × 4 m

Single plot size: 1.80 m × 1.96 m

Number of plants/plot: 15

Spacing: 0.4 m × 0.6 m

(plant to plant/row to row)

Variety: California Wonder

Factor A: Growing conditions

G_1: Naturally ventilated greenhouse

G_1: Naturally ventilated greenhouse

Factor B: Date of planting

S_1: September 2, 2009

S_2: September 21, 2009

S_3: October 12, 2009

9.3.7   BIOMETRIC OBSERVATIONS

Twelve experimental plots each of 1.80 m × 1.96 m were considered for both inside the ventilated greenhouse (G_1) and in open condition (control plot, G_2). The capsicum plants of variety California Wonder were planted. Each date of sowing (S_1,S_2,S_3) was replicated four times. The planting was done on September 2, 2009 (S_1), September 21, 2009 (S_2), and October 12, 2009

(S$_3$) both inside the greenhouse and in the control plots keeping the row to row and plant to plant spacing of 60 and 40 cm, respectively. The number of plants in each plot was 15 and the number of plants per square meter was four. During the experiment, the required dose of FYM (20 tons/ha) and chemical fertilizer (N:P:K :: 120:80:50 kg/ha) were applied. Irrigation was also applied as per the requirement of the plants [17–19]. During the growth period of the plants, pesticides were used depending upon the types of pests and diseases attacked to the plants. Four plants in each treatment were selected at random and utilized for recording the biometric parameters both for inside and outside the greenhouse.

9.3.7.1   PLANT HEIGHT

The average height of the plants both inside the greenhouse and in control plot were measured at 30, 60, and 90 days after planting (DAP) taking four numbers of randomly selected plants from each plot with the help of meter scale and expressed in centimeter.

9.3.7.2   NUMBER OF LEAVES PER PLANT

The average number of leaves per plant both inside the greenhouse and in control plot were measured at 30, 60, and 90 days after planting taking four numbers of randomly selected plants from each plot and expressed in number.

9.3.7.3   DAYS TO FIRST FLOWERING

The number of days taken for first flowering from the date of planting for both the plants inside the greenhouse and in control plot were recorded and expressed in days.

9.3.7.4   DAYS TO FIRST FRUITING

The number of days taken for first fruiting from the date of planting for both the plants inside the greenhouse and in control plot were recorded and expressed in days.

9.3.7.5   DAYS TO FIRST PLUCKING

The number of days taken for first plucking from the date of planting for both the plants inside the greenhouse and in control plot were recorded and expressed in days.

9.3.7.6   PERIOD OF HARVEST

The period of harvest from the first plucking to the last plucking for the crop both inside the greenhouse and in control plot was recorded and expressed in days.

9.3.7.7   NUMBER OF PLUCKING

The number of plucking for the crop both inside the greenhouse and in control plot was recorded and expressed in number.

9.3.7.8   DIAMETER OF FRUIT

The fruit diameter was measured by using the thread and the average diameter was expressed in centimeter for both the treatments.

9.3.7.9   LENGTH OF FRUIT

The length of fruit at harvest stage was measured with a meter scale and the average was taken which was expressed in centimeter for both the treatments.

9.3.7.10   WEIGHT OF FRUIT

Four fruits from each plant were randomly harvested and the average fruit weight was worked out and expressed in grams (g) for both treatments.

9.3.7.11   NUMBER OF FRUITS PER PLANT

The total number of fruits from each sample plant harvested during the harvest period were counted and recorded for both the treatments. Then the sum was calculated to get the number of fruits per plant.

9.3.7.12   FRESH FRUIT YIELD PER PLANT

The weight of fruits per plant from all the harvests were added and expressed in kilogram (kg) per plant.

9.3.7.13   FRUIT YIELD PER SQUARE METER

The yield of the fruits from the plants in square meter was recorded and expressed in kilogram (kg).

9.3.7.14   FRUIT YIELD PER HECTARE

The yield of fruits from the plant in one square meter was recorded and from that the fruit yield per hectare was obtained and expressed in tons/hectare (t/ha).

9.4   RESULTS AND DISCUSSION

This study was conducted to evaluate the effectiveness of greenhouse through cultivation of off-season vegetables. Various control parameters have accordingly been adjusted suitable for the better growth of capsicum inside the greenhouse after studying the thermal behavior of greenhouse through the model developed. The growth and yield of the capsicum inside the greenhouse has finally been compared with respect to the open field condition along with its cost of cultivation in both the cases.

9.4.1   GROWTH PARAMETERS

The growth parameters of capsicum at 30, 60, and 90 days after planting in different growing conditions, dates of planting and their interaction are presented below.

9.4.1.1   PLANT HEIGHT

The plant height of capsicum recorded after 30, 60, and 90 days after planting are presented in Table 9.1. The data analyzed for this trait indicated significant differences with respect to growing conditions irrespective of stages of observation.

At 30 DAP, Table 9.2 indicates that plant height was significant due to dates of planting and growing conditions. The maximum plant height of 24.87 cm and minimum of 14.44 cm was observed with G_1 and G_2, respectively irrespective of date of planting. Similarly higher plant height of 21.30 cm was recorded in the third date of sowing (S_3) which was at par with S_2. Minimum plant height was obtained with S_1 (17.63 cm). The interaction effect between growing condition and dates of planting was non-significant at 30 days after planting.

At 60 DAP, Table 9.2 shows significant differences for growing conditions, dates of planting on plant height at 60 days after planting. Highest plant height of 61.64, 44.10, and 64.13 cm were recorded in G_1 growing condition, S_3 date of planting and G_1S_3 interaction of growing condition, respectively. Lowest value was observed in G_2S_1 (19.34 cm) for this parameter. The interaction effect between growing condition and dates of planting was non-significant at 60 days after planting (Fig. 9.2).

TABLE 9.2   Effect of Growing Conditions, Dates of Planting and Their Interaction on Plant Height in cm.

Growing condition	Date of planting			Mean
	S_1	S_2	S_3	
	30 DAP			
G_1	21.85	25.91	26.87	24.87
G_2	13.42	14.16	15.74	14.44
Mean	17.63	20.03	21.30	
	G	S	G × S	
SE (m) ±	0.712	0.753	1.23	
CD (5%)	2.19	2.32	3.79	
	60 DAP			
G_1	58.41	62.40	64.13	61.64
G_2	19.34	23.18	24.07	22.19
Mean	38.87	42.79	44.10	

TABLE 9.2   *(Continued)*

Growing condition	Date of planting			Mean
	S_1	S_2	S_3	
	G	S	G × S	
SE (m) ±	1.10	1.162	1.90	
CD (5%)	3.39	3.58	5.84	
		90 DAP		
G_1	84.50	88.34	90.12	87.65
G_2	33.23	34.13	34.73	34.03
Mean	58.86	61.23	62.42	
	G	S	G × S	
SE (m) ±	0.740	0.782	1.28	
CD (5%)	2.28	2.41	3.93	

G_1 (Greenhouse) and G_2 (Open condition); S1 (date of planting September 2, 2009); S_2 (date of planting September 21, 2009) and S_3 (date of planting October 12, 2009)

At 90 DAP, Table 9.2 exhibits significant differences for growing condition on plant height at 90 DAP. Tallest plant measuring 87.65, 62.42, and 90.12 cm were recorded in G_1 growing condition, S_3 date of planting and G_1S_3 condition, respectively. Shortest plant was observed in G_2S_1 (33.23 cm). The interaction effect between growing condition and dates of planting was non-significant at 90 days after planting.

Maximum plant height under greenhouse was due to the shading effect resulting in lowering of temperature which was favorable for increasing the cell elongation thereby increased the length of the internodes leading to increase in plant height.

9.4.1.2   *NUMBER OF LEAVES PER PLANT*

The number of leaves per plant of capsicum recorded at 30, 60, and 90 days after planting is presented in Table 9.3, respectively. The data analyzed for this trait indicated significant difference with respect to growing condition irrespective of stages of observation.

At 30 DAP, Table 9.3 reveals higher number of leaves per plant (27.15) in G_1 where as it was lower (19.44) in G_2 irrespective of date of planting. Highest number of leaves per plant (29.10) was recorded in third date of planting (S_3), where as it was lowest (17.14) in S_1 irrespective of growing condition. There was no significant difference between S_2 and S_3 for growing conditions G_2 and G_1. The interaction effect between growing condition and dates of planting was non-significant at 30 days after planting.

TABLE 9.3 Effects of Growing Conditions, Dates of Planting and Their Interaction on Number of Leaves per Plant.

Growing condition	Date of planting			Mean
	S_1	S_2	S_3	
	30 DAP			
G_1	25.24	27.13	29.10	27.15
G_2	17.14	20.07	21.12	19.44
Mean	21.19	23.60	25.11	
	G	S	G × S	
SE (m) ±	0.872	0.992	1.51	
CD (5%)	2.69	2.84	4.64	
	60 DAP			
G_1	69.14	73.19	75.12	72.48
G_2	36.21	38.16	39.42	37.93
Mean	52.67	55.67	57.27	
	G	S	G × S	
SE (m) ±	1.271	1.344	2.194	
CD (5%)	3.92	4.14	6.76	
	90 DAP			
G_1	84.72	89.41	93.20	89.11
G_2	64.45	68.20	71.50	68.05
Mean	74.58	78.80	82.35	
	G	S	G × S	
SE (m) ±	1.02	1.08	1.77	
CD (5%)	3.16	3.34	5.45	

G_1 (Greenhouse) and G_2 (Open condition); S1 (date of planting September 2, 2009);
S_2 (date of planting September 21, 2009) and S_3 (date of planting October 12, 2009)

At 60 DAP, Table 9.3 reveals significant differences for growing conditions and dates of planting on number of leaves per plant at 60 days after planting. Highest number of leaves per plant of 72.48, 57.27, and 75.12 were recorded in G_1 growing condition, S_3 date of planting and G_1S_3 interaction of growing condition with dates of planting, respectively. Lowest value was observed in G_2S_1 for this observation. The interaction effect was non-significant.

At 90 DAP, Table 9.3 exhibits significant differences for growing conditions, dates of planting as well as their interaction for number of leaves per plant at 90 DAP. Highest numbers of leaves were 89.11, 82.35, and 93.20 in G_1 growing condition, S_3 date of planting and G_1S_3 condition, respectively. Lowest number of leaves was recorded in G_2S_1 (64.45). There was significant difference among dates of planting S_1, S_2, and S_3 for the growing condition G_2 and between G_1 and G_2 for all the dates of planting.

A significant increase in vegetative growth at different stages of the plant was observed in G_1 as compared to open field condition (G_2). The increase in trend of plant under greenhouse was due to the fact that the plants received comparatively lower light intensity than the plants under open conditions thereby facilitating cell elongation resulting in production of thicker plants. Also the congenial atmosphere created inside the greenhouse resulted in the increase of plant height. Increase in leaf number was the sum total of the carbohydrate production, which encouraged better growth of plant. Better environmental conditions provided under greenhouse condition resulted in more leaf production.

9.4.1.3   FRUIT DIAMETER

The yield attributing characteristics (fruit diameter, fruit length, fruit weight, number of fruits per plant, fruit yield per plant, yield per m², and fruit yield per hectare) for the capsicum grown in different growing conditions, different dates of planting and the interaction of both are presented in this section. The data analyzed in this trait indicated significant difference with respect to growing condition and different dates of planting. The data for these characteristics are presented in Tables 9.3 and 9.4, respectively.

Perusal of data on **fruit diameter** of capsicum presented in Table 9.3 exhibited wide range of variation from 2.51 cm in G_2 to 5.14 cm in G_1 for growing condition, 3.08 cm in S_1 to 4.00 cm in S_3 for dates of planting and 2.09 cm in G_2S_1 to 6.12 cm in G_1S_3 their interaction effect. The interaction effect between growing condition and dates of planting was non-significant on fruit diameter.

9.4.1.4   FRUIT LENGTH

Data on fruit length showed significant differences among growing conditions, dates of planting as well as their interaction effect in capsicum, which has been presented in Table 9.4. Perusal of the data exhibited variation ranging from 5.75 cm in G_1 to 2.94 cm in G_2 for growing conditions irrespective of dates of planting, where as it was lowest (3.36 cm) in S_1 irrespective of growing condition. Growing under naturally ventilated green house with third date of planting G_1S_3 produced the largest fruit having length 7.14 cm, where as it was lowest in G_2S_1 (2.09 cm) (Fig.9.3).

TABLE 9.4   Effects of Growing Conditions, Dates of Planting and Their Interaction on Fruit Diameter, Length, and Weight after 90 Days of Planting.

Growing condition	Date of planting			Mean
	S_1	S_2	S_3	
Fruit diameter in cm				
G_1	4.07	5.23	6.12	5.14
G_2	2.09	2.61	2.84	2.51
Mean	3.08	3.92	4.48	
	G	S	G × S	
SE (m) ±	0.390	0.412	0.673	
CD (5%)	1.20	1.27	2.07	
Fruit length in cm				
G_1	4.64	5.48	7.14	5.75
G_2	2.09	3.12	3.62	2.94
Mean	3.36	4.30	5.38	
	G	S	G × S	
SE (m) ±	0.534	0.565	0.923	
CD (5%)	1.646	1.74	2.84	
Fruit weight, g				
G_1	99.90	111.82	121.32	111.01
G_2	65.08	74.72	85.32	75.04
Mean	82.49	93.27	103.32	
	G	S	G × S	
SE (m) ±	2.74	2.90	4.73	
CD (5%)	8.44	8.92	9.59	
Number of fruits per plant				
G_1	7.27	9.12	10.13	8.84
G_2	4.10	5.07	5.34	4.83

TABLE 9.4   *(Continued)*

Growing condition	Date of planting			Mean
	S_1	S_2	S_3	
Mean	5.68	7.09	7.73	
	G	S	G × S	
SE (m) ±	0.541	0.571	0.932	
CD (5%)	1.66	1.76	2.87	

G_1 (Greenhouse) and G_2 (Open condition); S1 (date of planting September 2, 2009);
S_2 (date of planting September 21, 2009) and S_3 (date of planting October 12, 2009)

9.4.1.5   FRUIT WEIGHT

The data presented in Table 9.4 reveals significant variations for fruit weight under growing conditions, dates of planting with respect to fruit weight. Higher fruit weight (111.01 g) was observed in G_1 where as it was lower in G_2 (75.04 g) taking dates of planting together. Heaviest fruit (103.32 g) was recorded in S_3 where as it was lowest in S_1 (82.49 g) irrespective of growing condition. Fruit weight was maximum (121.32 g) in G_1S_3 followed by G_1S_2 and it was lowest in G_2S_1 (65.08 g). The interaction effect between growing condition and dates of planting was non-significant. The harvested fruit samples in open field condition and greenhouse condition have been shown in Figures 9.4 and 9.5, respectively.

FIGURE 9.2   Measurement of plant height.

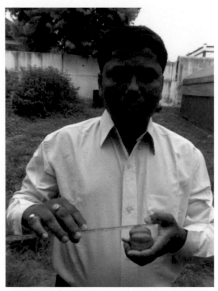

FIGURE 9.3   Measurement of fruit length.

FIGURE 9.4   Fruits harvested from open field cultivation.

FIGURE 9.5   Fruits harvested from greenhouse cultivation.

9.4.1.6   NUMBER OF FRUITS PER PLANT

Observations on number of fruits per plant in capsicum planted on different dates under different conditions are presented in Table 9.4. Significant differences were observed for growing conditions G_1 and G_2. The data in the Table revealed that capsicum grown under greenhouse (G_1) produced higher number of fruits (8.84) followed by open field condition G_2 (4.83), irrespective of dates of planting. Planting on third date (S_3) yielded highest number of fruits (7.73) followed by S_2 (7.09) and was lowest in S_1 (4.10) irrespective of growing condition. Planting of capsicum on the third date in the greenhouse (G_1S_2) produced highest number of fruits (10.13) per plant followed by G_1S_2 (9.12) and was lowest in G_2S_1 (4.10).

9.4.1.7   FRUIT YIELD PER PLANT

Significant variation with respect to green fruit yield per plant (kg) was observed under various growing conditions irrespective of the dates of planting. The dates are presented in Table 9.5. Data indicated higher fruit yield per plant (1.08 kg) in G_1 and lower (0.46 kg) in G_2 irrespective of dates of planting. Planting on third date (S_3) recorded higher fruit yield (0.77 kg) whereas lowest yield per plant (0.66 kg) was recorded in S_1 irrespective of growing conditions. Highest yield per plant (1.23 kg) was recorded in G_1S_3 followed by G_1S_2 (1.07 kg) and G_1S_1 (0.96 kg) in descending order. The interaction effect was non-significant.

9.4.1.8   FRUIT YIELD PER SQUARE METER (M²)

Significant difference was observed for this trait in different growing conditions as well as dates of planting which has been presented in Table 9.5. The yield per m^2 ranged in between 4.65 kg (G_1) and 1.98 kg (G_2) irrespective of different dates of planting. It varied from 3.79 kg (S_3) to 2.84 kg (S_1) in different dates of planting irrespective of growing conditions. Highest fruit yield was recorded in G_1S_3 (5.27 kg) followed by G_1S_2 (4.58 kg), G_1S_1 (4.11 kg). Lowest fruit yield per m^2 was recorded in open conditions ranging from 1.58 kg (G_2S_1) to 2.31 kg (G_2S_3). The interaction effect was non-significant.

9.4.1.9   FRUIT YIELD PER HECTARE

The data for fruit yield (tons/ha) for different growing conditions, planting dates and the interaction of both have been presented in Table 9.5. Fruit yield (t/ha) was higher (46.53 t) in G_1 followed by G_2 (19.80 t), irrespective of planting dates. There was a significant difference between the yields of the two growing conditions as well as the yield for different dates of planting. Highest fruit yield per hectare (37.90 t) was recorded in S_3, where as it was lowest in S_1 (28.45 t) irrespective of growing conditions. Highest fruit yield per hectare (52.70 t) was recorded in G_1S_3 followed by G_1S_2 (45.80 t) and G_1S_1 (41.10 t) and G_2S_3 (23.10 t) in descending order. Lowest value (15.80 t) was obtained in G_2S_1. The interaction effect was non-significant.

The above data revealed that the maximum yield (52.70 tons) of capsicum was obtained when it was planted inside the greenhouse during the month of October. It was 14% and 22% more than the yield obtained for the same growing condition with last week of September and first week of September planting. Also it gave 128.13% higher yield than the open field condition for October planting. Taking into account the two growing conditions, G_1 (Greenhouse cultivation) recorded 135% more yield than G_2 (Open field cultivation), irrespective of dates of planting.

TABLE 9.5   Effect of Growing Condition, Dates of Planting and Their Interaction on Fruit Yield after 90 Days of Planting.

Growing condition	90 days after planting			Mean
	S_1	S_2	S_3	
Fruit yield, kg per plant				
G_1	0.96	1.07	1.23	1.08
G_2	0.37	0.48	0.54	0.46
Mean	0.66	0.77	0.88	
	G	S	G × S	
SE (m) ±	0.028	0.029	0.047	
CD (5%)	0.085	0.09	0.146	
Fruit yield, kg per m2				
G_1	4.11	4.58	5.27	4.65
G_2	1.58	2.05	2.31	1.98
Mean	2.84	3.31	3.79	
	G	S	G × S	
SE (m) ±	0.120	0.127	0.278	
CD (5%)	0.37	0.39	0.86	
fruit yield per ha in tones				
G_1	41.1	45.8	52.70	46.53
G_2	15.80	20.50	23.10	19.80
Mean	28.45	33.15	37.90	
	G	S	G × S	
SE (m) ±	1.183	1.250	2.041	
CD (5%)	3.64	3.85	6.29	

G_1 (Greenhouse) and G_2 (Open condition); S1 (date of planting September 2, 2009); S_2 (date of planting September 21, 2009) and S_3 (date of planting October 12, 2009)

The parameters such as fruit length and diameter, number of fruits per plant, and fruit weight exhibited higher value for greenhouse condition in comparison to open condition. The day and night temperatures were higher in greenhouse with respect to open condition. In open conditions, the higher variation in day and night temperatures resulted in low value of these parameters. The important aspect that is, the yield which was very much related to the above-discussed parameters showed considerable variations under different growing conditions as well as under dates of planting. Fruit yield of capsicum in G_1 was about 2.35 times more than that obtained in open condition. Shading was also effective in lowering the temperature during peak sunny hours and in raising the temperature during cold night period inside the greenhouse thereby creating a better microclimate for production of quality fruits and higher yield. In general, higher yield was obtained in protected conditions than open conditions irrespective of planting time. This is in agreement with the findings for okra and capsicum [38], in tomato and eggplant [31] and in gerbera [46].

TABLE 9.6   Other Biometric Observations.

Observations	S_1		S_2		S_3	
	G_1	G_2	G_1	G_2	G_1	G_2
Days to first flowering	42	55	41	47	39	46
Days to first fruiting	55	63	53	59	50	56
Days to first plucking	85	96	81	89	78	88
Period of harvest	30	25	25	27	37	30
Number of plucking	05	03	06	03	06	03

G_1 (Greenhouse) and G_2 (Open condition); S1 (date of planting September 2, 2009); S_2 (date of planting September 21, 2009) and S_3 (date of planting October 12, 2009)

9.4.1.10   OTHER BIOMETRIC OBSERVATIONS

The biometric observations like days to flowering, days to fruiting, days to first plucking of fruit, period of harvest, and number of plucking for both the greenhouse and open conditions are presented in Table 9.6. These parameters play significant role toward the yield. The October planting of capsicum in greenhouse condition gave the best results from the agronomical point of view followed by the September planting under same conditions. The days to first flowering, days to first plucking from the date of planting were 39 and 50, respectively, for the October planting in greenhouse conditions. Also the other attributes like period of harvest, number of plucking were superior compared to other dates planting and conditions for which the yield was 52.70 t/ha followed by 45.8 t/ha obtained in greenhouse condition with last week of September planting. For all the three dates of planting, the greenhouse condition gave better result than the open filed condition.

9.4.2   COST BENEFIT RATIO (CBR)

The cost benefit ratio for growing capsicum inside the greenhouse came to be 2.98 (Table 9.7) whereas it was 0.80 when grown in open field condition (Table 9.7).

TABLE 9.7   Cost of Cultivation of Capsicum.

Particulars	Quantity	Rate (Rs.)	Amount (Rs.)
Inside the greenhouse (45 m² area)			
Seed	15 g.		30.00
Nursery management	1 man-day	100/man-day	100.00
Main field (land preparation)	3 man-day	100/man-day	300.00
FYM & fertilizer (urea) (55 kg/ha)	0.25 kg	15.00/kg	3.75
Single super phosphate (20 kg/ha)	0.09 kg	20.00/kg	1.80
Muriate of potash (30 kg/ha)	0.15 kg	20.00/kg	3.00
Compost preparation			70.00
Transplanting	1 man-day	100/man-day	100.00
Fertilizer application	1 man-day	100/man-day	100.00
Intercultural operations	3 man-day	100/man-day	300.00
Plant protection chemicals			200.00
Harvesting (10 harvests)	3 man-day	100/man-day	300.00
Miscellaneous expenses			500.00
Total cost of cultivation			2008.55
Average yield (kg/45 m²)	200 kg	40/kg	8000.00
Net return for 4 months, 8000 − 2008.55 =			5991.45
Benefit–cost ratio (5991.45/2008.55)			2.98
Outside the greenhouse (45 m² area)			
Seed	15 gm.		30.00
Nursery management	1 man-day	100/man-day	100.00
Main field (land preparation)	3 man-day	100/man-day	300.00
FYM & fertilizer (urea) (55 kg/ha)	0.25 kg	15.00/kg	3.75
Single super phosphate (20 kg/ha)	0.09 kg	20.00/kg	1.80
Muriate of potash (30 kg/ha)	0.15 kg	20.00/kg	3.00
Compost preparation			100.00
Transplanting	1 man-day	100/man-day	100.00

TABLE 9.7   *(Continued)*

Particulars	Quantity	Rate (Rs.)	Amount (Rs.)
Fertilizer application	1 man-day	100/man-day	100.00
Intercultural operations	3 man-day	100/man-day	300.00
Plant protection chemicals			370.00
Harvesting (8 harvests)	2 man-day	100/man-day	200.00
Miscellaneous expenses			500.00
Total cost of cultivation			2108.55
Average yield (kg/45 m²)	95 kg	40/kg	3800.00
Net return for 4 months, 380−2108.55 =			1691.45
Benefit–cost ratio (1691.45/2108.55)			0.80

9.5   CONCLUSIONS

The suitability of a low-tech naturally ventilated greenhouse was evaluated for off-season cultivation of bell pepper (capsicum) in coastal region of Odisha because of its high demand during that period. The cultivation of this vegetable was tried in winter days of 2009–2010 with three dates of sowings under different growing conditions. It was observed that the crop yield was better under greenhouse during off-season as compared to open field. The greenhouse with shade net proved to be the best modified protected condition for better plant growth and maximum yield as compared to without shade for less variation in temperature due to partial elimination of incoming radiation by the shade net during day hours and prevention of the radiative losses to the cold night sky thus maintaining a better heat distribution inside the greenhouse. Considering the greenhouse to be an effective solar energy collector, appropriate model was developed to study the thermal performance of the greenhouse for the above study. The developed model was then tested with the experimental results obtained which showed fair agreement for the heating mode. The model developed can also be used for parametric studies to improve the performance of the greenhouse. The economics of cultivating the vegetable inside the greenhouse justified its acceptability. As covering with UV sheet or UV sheet along with a shade net over a structure fulfill the requirement of a greenhouse, the small and marginal farmers of Odisha can grow off-season vegetables which is quite remunerative. On the basis of the above study, the following conclusions are drawn:

a. During peak sunny hours, the greenhouse air temperature inside the shade net was 2–3 °C higher than the ambient air temperature and in the night hours, the inside air temperature is 3–5 °C higher than the ambient air temperature.

b. The observed plant temperature inside the greenhouse with shade net during peak sunny hours is 1–2 °C lower than ambient air temperature and during night hours it is observed that, the plant temperature is 2–4 °C more than the ambient air temperature.

c. Natural ventilation was done (10 am to 4 pm) to keep the greenhouse air temperature within 3–4 °C lower than the ambient air temperature to make it suitable for the growth of the crops inside the greenhouse.

d. The variation of temperature is less in case of greenhouse with shade net than without shade net due to partial elimination of incoming radiation during sunny hours and preservation of the radiative heat losses to the cold night sky for maintaining better heat distribution inside the greenhouse during night hours due to shade.

e. Overall growth of capsicum in terms of height of plants and number of leaves per plant inside the greenhouse was better as compared to the open field. Early flowering and fruiting were observed in greenhouse condition. Plant height at 90 days after planting inside the greenhouse was 157.56% higher than open field condition.

f. The fruit length inside the greenhouse was 95.57 % higher than open field condition.

g. The weight of single fruit was 47% higher than the same for open field condition.

h. The yield of capsicum per square meter of the cultivated area was 2.34 times more than open field condition irrespective of the date of planting. The same for October planting was observed to be 1.28 and 1.15 times more than the planting during first week of September and last week of September. Also for October planting, the yield (46.53 t/ha) inside the greenhouse was 2.28 times more than open field condition (19.80 t/ha).

i. The benefit cost ratio for capsicum in the greenhouse was 2.98:1 whereas it was 0.80:1 in case of open field condition.

j. Considering the yield, environmental parameters and cost of cultivation, the naturally ventilated greenhouse with shade net is suitable for the cultivation of capsicum during off-season in coastal Odisha.

KEYWORDS

- bell pepper
- benefit cost ratio
- biometric parameters
- capsicum
- energy
- conservation
- flowering
- fruiting
- greenhouse
- growth parameters
- India
- plant growth
- radiative losses
- shade net
- solar energy
- solar still
- thermal insulation
- tomato
- vegetable cultivation
- vegetables
- water use efficiency

REFERENCES

1. Ali, I. A.; Abdullah, A. M. Microcomputer Based System for All Year Round Temperature Control in Greenhouse in Dry Arid Lands. *Comput. Electron. Agric.* **1993,** *8*(3), 195–210.
2. Al-Amri, Ali. M.S. Solar Energy Utilization in Greenhouse Tomato Production. *J. King Saudi Univ. Agric. Sci.* **1997,** *9*, 21–38.
3. Anonymous. *Hand Book of Agriculture*; ICAR Publication: New Delhi, India, 1992; pp 93.
4. Anonymous. *Economic Survey*; Directorate of Economics and Statistics, Planning and Coordination Department: Government of Odisha, 2000.

5. Arinze, E. A.; Schoenau, G. J.; Besant, R. W. Experimental and Computer Performance Evaluation of a Movable Thermal Insulation for Energy Conservation in Greenhouses. *J. Agric. Eng. Res.* 1986, *34*, 97–113.

6. ASHRAE. *Fundamental Handbook Environmental Control for Animals and Plants*; ASHRAE: New York, 1997; Chapter 10, pp 17.

7. Bakker, J. C. Effects of Day and Night Humidity on Yield and Fruit Quality of Glasshouse Tomatoes. *J. Hort. Sci.* **1990**, *65*, 324.

8. Bargach, M. N.; Tadili, R.; Dahman, A. S.; Boukallouch, M. Survey of Thermal Performances of a Solar System Used for the Heating of Agricultural Greenhouses in Morocco. *Renew. Energ.* **2000**, *20*, 415–433.

9. Barral, J. R.; Galimberti, P. D.; Barone, A.; Miguel, A. L. Integrated Thermal Improvements for Greenhouse Cultivation in the Central Part of Argentina. *Sol. Energy.* **1999**, *67*(1–3), 111–118.

10. Boulard, T.; Kittas, C.; Roy, J. C.; Wang, S. Convective and Ventilation Transfers in a Greenhouse. *Biosyst. Eng.* **2002**, *83*(2), 129–147.

11. Chandra, P.; Albright, L. D. Analytical Determination of the Effect on Greenhouse Heating Requirements of Using Night Curtains. *T. ASAE.* 1980, *23*(4), 994–1000.

12. Cooper, P. I. Maximum Efficiency of a Single Effect Solar Still. *Sol. Energ.* **1973**, *15*, 205.

13. Critten, D. L. Optimization of CO_2 Concentration of a Greenhouse: A Modeling Analysis for the Lettuce Crop. *J. Agric. Eng. Res.* **1991**, *48*, 261–271.

14. Fernandez, J.; Chargoy, N. Multistage Indirectly Heated Solar Still. *Sol. Energ.* **1990**, *44*(4), 215.

15. Ganeshan, M. Performance of Tomato (*Lycopersicon Esculentum Mill.*) Varieties under Organic Farming in Greenhouse and Open Field Conditions during Winter Season of Tamil Nadu, Research Notes. *Madras Agric. J.* **2001**, *88*(10–12), 726–727.

16. Garzoli, K.; Blackwell, J. An Analysis of the Nocturnal Heat Loss from a Single Skin Plastic Greenhouse. *J. Agric. Eng. Res.* 1981, *26*, 203–214.

17. Goyal, M. R. Evapotranspiration. In *Principles and Applications for Water Management*; Harmsen, E. W., Ed.; Apple Academic Press Inc: Oakville, ON, 2014.

18. Goyal, M. R. *Research Advances in Sustainable Micro Irrigation;* Apple Academic Press Inc: Oakville, ON, 2015; Vol. 1–10.

19. Goyal, M. R. *Innovations and Challenges in Micro Irrigation;* Apple Academic Press Inc: Oakville, ON, 2016; Vol. 1–4.

20. Grafiadellis, M. *In Greenhouse Heating With Solar Energy*; Von Zabeltitz, C., Ed.; FAO: Rome, Italy, 1987; pp 89–93.

21. Grafiadellis, G. A.; Kyritsis, S. Heating Greenhouses with Solar Energy. *Acta. Hort.* 1981, *115*, 553–564.

22. Gupta, M. J.; Chandra, P. Effect of Greenhouse Design Parameters on Conservation of Energy for Greenhouse Environment Control. *Energy.* 2002, *27*, 777–794.

23. Gupta, A.; Tiwari, G. N. Computer Model and Its Validation for Prediction of Storage Effect of Water Mass in a Greenhouse: A Transient Analysis. *Energ. Convers. Manage.* **2002**, *43*(18), 2625–2640.

24. Hanan, J. J.; Holley, W. D.; Goldsberry, K. L. *Greenhouse Management*; Springer Verlag: Berlin, 1978; pp 231.

25. Holder, R.; Cockshull, K. E. Effects of Humidity on the Growth and Yield of Glasshouse Tomatoes. *J. Hort. Sci.* **1990**, *65*, 33.

26. Hussaini, H.; Suen, K. O. Using Shallow Solar Ponds as a Heating Source for Greenhouses in Cold Climate. *Energ. Convers. Manage.* **1998,** *39*(13), 1369–1376.

27. Jain, D.; Tiwari, G. N. Modeling and Optimal Design of Ground Air Collector for Heating in Controlled Environment Greenhouse. *Energ. Convers. Manage.* **2003,** *44*(8), 1357–1372.

28. Jensen, M. H.; Malter, A. J. *Protected Agriculture-: A Global Review*; The World Bank Technical Paper: Washington DC, 2005; Vol. 253, pp157.

29. Jolliet, O. H. A Model for Predicting and Optimizing Humidity and Transpiration in Greenhouse. *J. Agric. Eng. Res.* **1994,** *57*, 23–37.

30. Kantaswamy, V.; Singh, N.; Veeraragavathanatham, D.; Srinivasan, K.; Thiruvudainambi, S. Studies on Growth and Yield of Cucumber and Sprouting Broccoli under Polyhouse Condition. *South Indian Hort.* **2002,** *48*(1–6), 47–52.

31. Kothari, S. Thermal Modeling and Performance Evaluation of Greenhouse. Unpublished Ph.D. Thesis, Indian Institute of Technology: Delhi, 1999.

32. Kumar, R.; Srivastava, B. K. Effect of Plastic Covering on the Growth of Winter Grown Tomatoes under Low Plastic Tunnels. *Indian J. Agric. Res.* **2002,** *36*(4), 278–281.

33. Kurata, K. Simulation of Inside Air Temperature, Humidity and Crop Production in an Energy Conservation Greenhouse. *Acta. Hort.* **1989,** *245*, 339–345.

34. Kurpaska, S.; Slipek, Z. Optimization of Greenhouse Substrate Heating. *J. Agric. Eng. Res.* 2000, *76*, 129–139.

35. Ghosal, M. K.; Tiwari, G. N. Off-Season Cultivation of Capsicums in a Solar Greenhouse. *Int. J. Ambient Energ.* **2001,** *22*(4), 189–198.

36. Nayak, S. C.; Uttaray, S. K.; Dash, A. N.; Lenka, D. Use of Plastic Tunnels for Paddy Straw Mushroom Cultivation and Raising Seedlings. In *The Use of Plastics in Agriculture, Proceedings of XI International Congress,* Feb 26, March 2, 1990; New Delhi, India, 1990, E135–140.

37. Nelson, P. V. *Greenhouse Operation and Management.* Reston Publishing Company Inc.: Reston, VA, 2005; pp 347–362.

38. Nimje, P. M.; Wanjari, O. D.; Shyam, M. Greenhouse Technology for Vegetable Crop Production. In *The Use of Plastics in Agriculture, Proceedings of XI International Congress*, Feb 26, March 2, 1990; New Delhi, India, 1990, E178–182.

39. Patel, B. R.; Gohil, K. B.; Vaghasiya, P. M. Response of Gladiator Rose to Greenhouse under Varied Planting Geometry. *Agric. Eng. Today.* **2003,** *27*(3–4), 49–54.

40. Patel, B. R.; Raman, S. Effect of Low Tunnel on Growth Performance of Winter Paddy Seedlings. *Agric. Eng. Today.* **2002,** *26*(5–6), 45–50.

41. Plaza, S. D. L.; Benavente, R. M.; Garcia, J. L.; Navas, L. M.; Luna, L, Duran, J. M.; Retamal, N. Modeling and Optimal Design of an Electric Substrate Heating System for Greenhouse Crops. *J. Agric. Eng. Res.* **1999,** *73*(2), 131–139.

42. Sethi, V. P.; Lal, T.; Gupta, Y. P.; Hans, V. S. Effect of Greenhouse Micro-Climate on the Selected Summer Vegetable. *J. Res. Punjab Agric. Univ.* **2003,** *40*(3–4), 415–419.

43. Sharma, P. K. Computer Modeling and Experimental Testing of Greenhouse for Off-Season Crop Production for Rural Application. Unpublished Ph.D. Thesis, Indian Institute of Technology: Delhi, 1998.

44. Sharma, P. K.; Tiwari, G. N.; Soryan, V. P. S. Parametric Studies of a Greenhouse for Summer Conditions. *Energ.* **1998,** *27*, 733–740.

45. Shujuan, Z.; Yong, H.; Shuangxi, W. The Evaluation of Design Quality of a Sunlight Greenhouse Based on the Fuzzy Method. *Trans. Chin. Soc. Agric. Eng.* **2002,** *33*(5), 67–70.

46. Singh, R. D.; Tiwari, G. N. Thermal Heating of Controlled Environment Greenhouse: A Transient Analysis. *Energ. Conserv. Manage.* **2000,** *41,* 505–522.
47. Sirjacobs, M. *Greenhouses in Egypt, Protected Cultivation in the Mediterranean Climate;* FAO: Rome, Italy, 1989.
48. Verlodt, H. *Greenhouses in Cyprus, Protected Cultivation in the Mediterranean Climate*; FAO: Rome, Italy, 1990.

APPENDIX – I   WEEKLY AVERAGE AIR TEMPERATURE, RELATIVE HUMIDITY, LIGHT INTENSITY, AND SOIL TEMPERATURE AT 7:30 A.M., DURING 2009–2010.

SMW	Temperature, °C		Relative humidity %		Light intensity, lux	
	GH	Open	GH	Open	GH	Open
41	32.2	31.62	66.3	68.6	1775	1825
42	31.19	30.2	67.1	70	1800	2000
43	30.14	29.9	66.8	69.3	1735	1950
44	30.23	29.41	66.2	68.4	1650	1800
45	29.6	28.5	65.3	68.1	1550	1850
46	28.67	27.56	60.4	62.6	1500	1750
47	28.83	27.9	55.6	57.2	1450	2100
48	28.63	27.78	57.3	60.7	1350	1875
49	28.13	27.08	55.6	60.3	1200	2175
50	24.8	24.04	67.33	57.1	1025	1625
51	25.4	24.9	58.5	56.2	1250	2050
52	25.17	24.67	60.3	53.1	1333	1433.3
1	23.65	23.25	63.2	48.2	1175	1700
2	22.5	22.15	59.1	42.1	1217	1950
3	22.84	22.29	73.6	38.4	1500	1900
4	21.8	21.16	72.2	43	1100	1800
5	21.9	21.06	71.6	46.3	1250	1800
6	21.5	20.41	71	52.4	1050	1800
7	24.55	23.39	67.5	56.1	1125	2525
8	25.16	24.07	68.3	64.6	1233	1530
9	26.6	25.46	63.5	62.5	1625	2000
10	27.8	26.34	64.1	62.1	1560	1780
11	28.13	26.82	65.2	63	1600	1853
12	28.91	27.16	66.1	62.2	1750	1910
13	29.13	28.22	67	63.1	1800	2000

APPENDIX – II   WEEKLY AVERAGE AIR TEMPERATURE, RELATIVE HUMIDITY, LIGHT INTENSITY, AND SOIL TEMPERATURE AT 12:30 P.M., DURING 2009–2010.

SMW	Temperature, °C		Relative humidity %		Light intensity, lux	
	GH	Open	GH	Open	GH	Open
41	41.24	35.19	53.2	36	3500	4025
42	40.93	34.82	53.6	42.2	3415	3975
43	40.13	33.91	52.4	40.1	3385	3865
44	39.82	33.28	51.3	38.2	3475	3925
45	39.97	33.17	55	43.2	3550	3875
46	38.3	31.85	55.1	34.3	3250	3800
47	38.5	33.13	53.2	30.7	3100	3750
48	35.94	30.57	56.6	44.3	3000	3825
49	35.67	30.44	44	38	2700	3466.6
50	34.2	2831	58.1	47.3	2275	2675
51	35.9	31.4	54.3	34.5	3000	4200
52	34.27	31.83	57.3	43.3	2866	3900
1	33.9	30.1	59.1	33.1	2350	3875
2	34.25	31.75	48.3	39.6	2750	4325
3	35.25	31.55	54.1	40.5	3500	4500
4	32.72	27.3	55.5	53.8	2700	3750
5	32.8	26.68	54.3	46.4	2500	4200
6	33.6	27.62	50.8	47.1	2450	3850
7	35.95	30.74	56	48.4	3075	4350
8	35.84	29.18	47.9	43.5	3025	4450
9	36.25	31.59	50.5	45	3200	4450
10	36.81	32.43	50.2	46.1	3275	4325
11	37.23	32.89	51.8	47	3386	4415
12	38.14	33.44	52.3	47.8	3419	4460
13	38.62	34.19	55.2	48.1	3525	4530

APPENDIX – III   WEEKLY AVERAGE AIR TEMPERATURE, RELATIVE HUMIDITY, LIGHT INTENSITY, AND SOIL TEMPERATURE AT 4:30 P.M.

SMW	Temperature, °C		Relative humidity %		Light intensity, lux	
	GH	Open	GH	Open	GH	Open
41	31.15	30.43	63.9	62.4	630	710
42	31.26	29.16	64.2	61.8	575	725
43	31	28.28	65.4	62.3	680	760
44	30.48	27.14	62.3	59.1	675	720
45	30.27	26.99	63.1	58	650	750
46	27.13	23.86	60.8	56.3	425	525
47	24.8	21.59	57.3	48.6	350	375
48	26.24	22.58	59.6	49.7	400	450
49	25.37	22.36	56.4	47.6	183	225
50	25.15	22.08	62.5	57.6	175	325
51	25.35	23.18	60.3	43.1	300	350
52	25.34	22.49	57.6	48.7	316	367
1	24.55	22.03	42.5	35.2	275	350
2	24.5	21.56	52.7	42.5	300	575
3	25.13	22.26	51.3	40.3	300	750
4	26.46	23.41	61.8	50.5	250	625
5	26.2	23.12	59.4	54.4	400	500
6	25.88	22.19	58.3	54.2	350	450
7	27.2	23.51	59.7	49.2	475	950
8	27.56	23.63	62.3	50	550	875
9	29.78	25.93	69.2	58.4	550	1050
10	29.65	26.29	68.4	57.3	600	995
11	29.93	27.14	69.1	58.4	575	1025
12	30.14	28.62	68.2	58.8	590	1075
13	31.28	29.18	70.9	62.6	650	1215

CHAPTER 10

ADAPTABILITY OF DRIP IRRIGATED CANOLA IN DIFFERENT REGIONS OF EGYPT

MONA A. M. SOLIMAN[1], N. M. MAHROUS[2], SABREEN K. H. PIBARS[3], and M. HASSIB[1]

[1]Department of Water Relation and Field Irrigation, National Research Center, Dokki, Giza, Egypt

[2]Department of Agronomy, Faculty of Agriculture, Cairo University, Cairo, Giza, Egypt

[3]Department Water Relations and Field Irrigation, Agricultural and Biological Division, National Research Center, Cairo, Giza, Egypt

CONTENTS

ABSTRACT

Salinity is a major abiotic stress that significantly influences the germination and growth of plants. The objectives of this study were: to compare different types of soils and water quality (loam soil + low salinity irrigation water, sand silt soils + medium salinity irrigation water and saline soil + high salinity irrigation water) on the seed yield and seeds per pod of four genotypes of canola (AD201, Semu 84/204, Canola 101, and H4). The experiment was laid in split plot randomized complete block design (RCBD) with three replications. The results of this experiment showed that as the salinity stress intensifies the seed yield of all genotypes was decreased. Significant mean square of the salinity levels, genotypes, and salinity × genotypes interaction effects were exhibited for seeds per pod and seed yield. In all varieties, this interaction can be put in the following ascending order: Fayoum × genotypes < 60 km Desert Road Cairo × genotypes < El-Sharkia × genotypes. The differences in the studied parameters between any two interactions were significant at the 5% level.

10.1   INTRODUCTION

Salinity is a major problem that negatively impacts agricultural activities in many regions in the world, and especially when the source of irrigation water is limited the Near East and North Africa region. Generally, salinity problems increase with increasing salt concentration in irrigation water. Crop growth reduction due to salinity is generally related to the osmotic potential of the root-zone soil solution. This will lead to certain phonological changes and substantial reduction in productivity. Salinity also, affects the soil physical properties.

Salinity is a well-known problem in most or all arid and semi-arid regions of the world especially in irrigated areas. Salinity limits productivity of irrigated soils in vast areas of the world [4, 12]. Over 400 Mha across the world is affected by either salinity or sodicity, which accounts for about 6% of the world's land. Of the current 230 Mha of irrigated land in the world, 45 Mha are salt-affected (19.5%), and of the 1500 Mha under dry land agriculture 32 Mha (2.1%) are salt-affected to varying degrees [7].

Salinity, drought, nutrient imbalances, and temperature extremes are among the chief abiotic stresses impairing crop productivity worldwide [6, 9]. Although, salinization occurs mostly in arid- and semi-arid regions of the world but in reality no climatic zone is free from salinity [3, 17]. Salinity hampers the crop growth by creating low osmotic potential of soil solution (water stress), nutritional imbalance, specific ion effect (salt stress), and/or combination of these factors [1, 14]. Decrease in osmotic potential is linked with the accumulation of ions in soil solution, whereas nutritional imbalance and specific ion effect is linked with the higher accumulation of ions mainly Na+ and Cl− at toxic levels which leads to lessen the absorption availability of other essential elements like potassium and calcium etc. [5].

Canola is an important oilseed with worldwide importance (Appendix–I). It is currently ranked third, after soybean and palm oils, and fifth in the world trade in agricultural crops, after rice, wheat, maize, and cotton. In an effort to develop the low erucic acid cultivars, the Canadian plant breeders also attempted to lower the glucosinolate content of the oil-free seed meal. The meal is an excellent source of protein with a favorable balance of amino acids; however, the glucosinolates, which can cause nutritional problems, limit the use of the seed meal as a supplemental animal feed [18].This intensive breeding program resulted in Canada becoming the first country to produce rapeseed cultivars with low erucic acid in the oil and low glucosinolates in the meal. To differentiate between these double-low cultivars and other rapeseed cultivars, the double-lows were called canola.

The increasing awareness of the health advantages of canola oil, which contains < 70 g kg^{-1} saturated fat, will undoubtedly result in an increasing demand for this product. This demand, as well as the search for alternative crops by growers, may result in plantings on soils where salinity problems already exist or may develop from the use of saline irrigation water. Although a few preliminary studies on the salt tolerance of rapeseed have been conducted in small pot cultures [2, 15], salt tolerance data are not available to predict canola yield responses in the field.

A few studies were conducted about the effects of salinity on canola. He and Cramer [11] investigated the effects of seawater salinity on six *Brassica* species and reported that *Brassica napus* (canola) was the most tolerant to salinity among the other species, such as *Brassica campestris, Brassica juncea, Brassicacarinata, Brassica nigra, and Brassica oleracea.* Redmann et al. [16] evaluated the seedling emergence and plant growth of two canola cultivars, HCN92 and Legend, in response to soil salinity under growth chamber conditions in Canada. Salinities varied between 0.8 and 11 ds/m. Salinity increased significantly reducing total seedling emergence and emergence rate, and also caused decreased leaf area, shoot and root biomass and evapotranspiration for both cultivars. Huang and Redmann [13] reported that *Brassica napus* was more salt tolerant than *Brassica kaber* based on the growth responses like total dry matter. In another study, Ashraf and McNeilly [2] reported that *Brassica napus* and *Brassica carinata* were considered relatively salt tolerant, whereas *Brassica campestris* and *Brassica juncea* relatively sensitive to salinity. Goyal has reported benefits of drip irrigation in several crops and dry regions [8].

The objectives of this study were to the effect of water salinity on seed yield, and seed yield per pod of some canola genotypes and their adaptability to the soil and climatic conditions of different regions of Egypt.

10.2 MATERIALS AND METHODS

10.2.1 EXPERIMENTAL LAYOUT

During winter season, a research study was conducted on four genotypes of canola (AD201, Semu 84/204, Canola 101, and H4) and in three regions:

a. El- Sharkia region {loamy soil+ low salinity irrigation water (0.4 ds/m)},
b. Cairo–Alexandria Desert Road region (60 km), (sandy silt soils + medium salinity irrigation water), and
c. Fayoum region (saline sandy soil + high salinity irrigation water).

The physical and chemical properties of soil and water are presented in Tables 10.1 and 10.2. Field experiments were conducted during two consecutive growing seasons in a randomized completely block design (RCBD) with four replications. Four seeds of each genotype were sown at 15 cm apart between hills and 50 cm between rows. The plots were fertilized with 25 kg of N/fed. in the form of urea, 30 kg of P_2O_5 and 50 kg of K_2O in the form of single super phosphate and potassium sulphate. The single super phosphate and 50% of potassium sulphate were applied at planning time. The nitrogen and the rest of potassium need were applied equally in five applications with five irrigations after germination. The crop was grown under drip irrigation.

TABLE 10.1 Some Soil Chemical and Physical Properties of the Experimental Site.

Region	pH 1:2.5	ECeds/m 1:5	Soluble cations, meq/L				Soluble anions, meq/L				Sand	Silt	Clay	Texture class
			Ca^{++}	Mg^{++}	Na^+	K^+	CO_3^-	HCO_3^-	SO_4^-	Cl^-				
Fayoum	8.4	6.82	9	5	64	0.41	8.1	2.4	29	39	72	14	14	SSL
60 km Misr Alex.	7.8	0.78	2.8	0.33	3	0.52	1.3	1.06	2.6	1.85	77.8	8.4	13.8	SL
El-Sharkia	7.82	1.11	3.84	1.69	4.9	0.59	1.52	4.11	3.71	1.71	21.6	29.3	49.1	Loamy

TABLE 10.2 Some Chemical and Physical Data of the Irrigation Water.

Region	pH	EC ds/m	Soluble cations, meq/L				Soluble anions, meq/L			CO_3^{-ppm}
			Ca^{++}	Mg^{++}	Na^+	K^+	HCO_3^-	SO_4^-	Cl^-	
Fayoum	7.4	5.11	3.2	5	49.1	0.22	3.2	21.5	32.2	3270.4
60 km Misr Alex.	7.43	2.5	2.01	0.63	1.94	–	4.01	0.91	1.82	403.2
El-Sharkia	7.46	0.4	1.88	2.1	1.87	–	4.11	2.01	1.3	345.6

10.2.2   IRRIGATION METHOD

Surface drip irrigation method was used. Standard drippers were spaced 0.3 m apart along 50 m lateral. Dripper discharge was 4 lph.

10.2.3   POTENTIAL EVAPOTRANSPIRATION IRRIGATION INTERVALS AND IRRIGATION WATER REQUIREMENTS

Potential Evapotranspiration (ET_p, mm day^{-1}) was calculated using the Eq 10.1 by Hargreaves and Samani [10]. Irrigation interval (I, days) is defined in Eq 10.2. Irrigation requirements (IR, m^3/fed.) are defined in Eq 10.3.

$$ET_p = [0.0075 \times TF \times SS \times KS \times ETR] \tag{10.1}$$

$$I = [(A.W \times A.D \times Rd) / ET_c] \times Ei \tag{10.2}$$

$$IR = ET_c \times I\,(\,1 + LR) \times 4.2 \tag{10.3}$$

In Eq 10.1

ET_p	= Potential evapotranspiration	(mm/day)
TF	= Mean daily temperature, TF= $(100 \times n/N)\ 0.5$	(°F)
SS	= Sunshine coefficient	
N	= Mean daily duration of max. possible sunshine hours	(hours)
n	= Actual mean daily duration of sunshine,	(hours)
KS	= Solar radiation coefficient = KS = $0.097 - 0.00042 \times RH$	
RH	= Mean daily relative humidity	(%)
ETR	= Extraterrestrial radiation	(mm/day)

In Eq 10.2

I	= Allowable intervals between two irrigation,	(day)
AW	= Available soil water, Aw = FC − PW	(mm/m)
FC	= Field capacity	(mm)
PW	= Permanent wilting point	(mm)
AD	= Allowable soil moisture depletion below field capacity	(%)
Rd	= Rooting depth	(cm)

ET_c = Actual evapotranspiration, $ET_c = ET_a \times K_c$,   (mm/day)

K_c = Crop coefficient

Ei = Irrigation efficiency   (%)

In Eq 10.3

IR = Irrigation requirement   (m³/fed.)

LR = Leaching requirement $\{EC_{IW}/2\ EC_{DW}\}$ and EC_{IW},   (ds/m)
and EC_{DW} = EC of irrigation water and drainage water, respectively, in ds/m.

10.2.4 TOTAL YIELD

The total yield for each treatment was determined using a frame of $1\times\ 1\ m^2$ size. The frame was placed randomly. The seed and straw of canola plants within the frame were weighed.

10.3 RESULTS AND DISCUSSION

10.3.1 MAIN EFFECTS OF VARIETY

The Figures 10.1 and 10.2 show that genotypes of canola were significantly different in all regions. The highest seed yield was obtained from AD201 genotype (1204.2, 1120.3, and 479.1 kg/fed.) in El-Sharkia, 60 km desert road Cairo–Alexandria and El-Fayoum region, respectively. The genotypeH4 produced the lowest seed yield (303.4 kg/fed.) in Fayoum region.

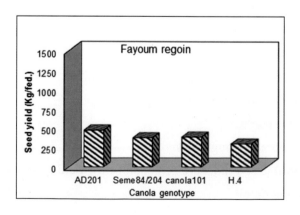

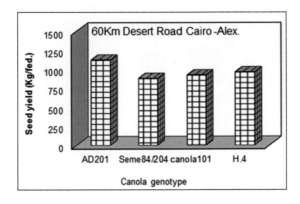

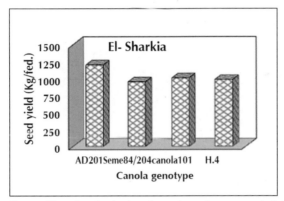

FIGURE 10.1   Effects of location (region) and canola variety on seed yield (kg/fed.).

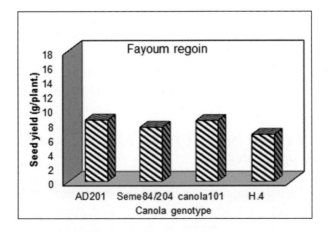

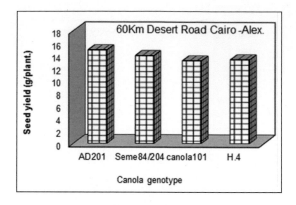

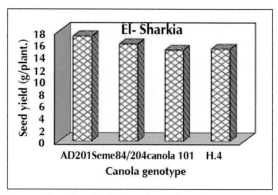

FIGURE 10.2   Effects of location (region) and canola variety on seed yield (g/plant).

10.3.2   MAIN EFFECT OF LOCATIONS (REGIONS)

Results showed that only the locations (regions) were significantly different for this character. El- Sharkia had the highest mean seed yield of 1204.2 kg/fed., because of its favorable climatic conditions as compared to other regions. Fayoum had the lowest seed yield (392.11 kg/fed.).

10.3.3   EFFECTS OF INTERACTION: CANOLA VARIETY × LOCATIONS (REGIONS)

In all varieties, this interaction can be put in the following ascending order: (Fayoum × variety) < (60 km Desert Road Cairo, Alexandria × variety) < (El-Sharkia × variety). Differences in the studied parameters between any two interactions were significant at the 5% level. The maximum values were obtained with the interaction (AD201 × El-Sharkia), whereas the minimum was with the interaction (H4 × Fayoum) except water use efficiency (WUE).

10.4   CONCLUSIONS

a.   Under farming conditions in the loamy soil and low salinity irrigation water, the order of variety in terms of preference for agriculture is as follows:
Seme 84/204 < H4 < Canola101 < AD201 hybrid

b.   Under the conditions of agriculture in sandy silty soils and medium salinity irrigation water, variety for the best seed yield followed the following order:
Seme 84/204 < Canola101 < H4 < AD201

c.   Under the conditions of agriculture in saline soil and high salinity irrigation water, variety in terms of preference is as follows:
H4 < Seme 84/204 < Canola101 < AD201

KEYWORDS

- **agro-climatic regions**
- **canola**
- **drip irrigation**
- **Egypt**
- **feddan**
- **irrigation interval**
- **irrigation requirement**
- **potential evapotranspiration**
- **productivity**
- **soil texture**
- **water saline**

REFERENCES

1. Ashraf, M. Breeding for Salinity Tolerance in Plants. *Crit. Rev. Plant Sci*. **1994**, *13*, 17–42.
2. Ashraf, M.; Mcneilly, T. Responses of Four *Brassica* Species to Sodium Chloride. *Environ. Exp. Bot*. **1990**, *30*, 475–487.
3. Bhutta, W. M.; Ibrahim, M.; Akhtar, J.; Shahzad, A.; Haq, T. U.; Haq, M. A. U. Comparative Performance of Sunflower (*Helianthus Annuus* L.) Genotypes against Nacl Salinity. *Santa Cruz Do Sul*. **2004**, *16*, 7–18.
4. Corwin, D. L.; Rhoades, J. D.; Vaughan, P. J. GIS Applications to the Basin-Scale Assessment of Soil Salinity and Salt Loading to Groundwater. *SSSA. Spec .Publ*. **1996**, *48*, 295–313.
5. El-Bassiouny, H. M. S.; Bekheta, M.A. Role of Putrescine on Growth, Regulation of Stomatal Aperture, Ionic Contents and Yield by Two Wheat Cultivars under Salinity Stress. *Egyptian J. Physiol. Sci*. **2001**, *2*, 239–258.
6. Farooq, M.; Bramley, H.; Palta, J. A.; Siddique, K. H. M. Heat Stress in Wheat during Reproductive and Grain Filling Phases. *Crit. Rev. Plant Sci*. **2011**, *30*, 491–507.
7. Ghassemi, F.; Jakeman, A. J.; Nik, H. A. *Salinization of Land and Water Resources*; University of New South Wales Press: Sydney, 1995; pp 526.
8. Goyal, M. R. *Research Advances in Sustainable Micro Irrigation;* Apple Academic Press Inc: Oakville, ON, Canada, 2015; Vol. 1–10.
9. Hamida, M. A.; Shaddad, M. A. K. Salt Tolerance of Crop Plants. *J. Stress Physiol. Biochem*. **2010**, *6*, 64–90.
10. Hargreaves, G. H.; Samani, Z. A. Estimating Potential Evapotranspiration. *J. Irrig. Drain.(ASCE)*. **1982**, *108*(3)*, 223–230.
11. He, T.; Cramer, G. R. Growth and Mineral Nutrition of Six Rapid-Cycling Brassica Species in Response to Seawater Salinity. *Plant Soil*. **1992**, *139*, 285–294.
12. Homaee; Feddes, M. R. A.; Dirksen, C. A Macroscopic Water Extraction Model for Nonuniform Transient Salinity and Water Stress. *Soil Scisoc. Am. J*. **2002**, *66*, 1764–1772.
13. Huang; Redmann, J. R. E. Physiological Responses of Canola and Wild Mustard to Salinity and Contrasting Calcium Supply. *J Plant Nutr*. **1995**, *18*, 1931–1949.
14. Marschner, H. *Mineral Nutrition of Higher Plants*; Academic Press: London, 1995.
15. Munshi, S. K.; Bhatia, N.; Dhillon, K. S.; Sukhija, P. S. Effect of Moisture and Salt Stress on Oil Filling in Brassica Seeds. *Proc. Indian Natl. Sci. Acad*. **1986**, B*52*, 755–759.
16. Redmann, R. E.; Qi, M. Q.; Belyk, M. Growth of Transgenetic and Standard Canola (B. Napus) Varieties to Soil Salinity. *Can. J. Plant Sci*. **1994**, *74*, 797–799.
17. Rengasamy, P.; World Salinization with Emphasis on Australia. *J. Exp. Bot*. **2006**, *57*, 1017–1023.
18. Thomas, P. Canadian Canola Production. In *Pacific Northwest Winter Rapeseed Production Conf.;* Moscow, ID, USA, 24–26 Feb, 1986; Kephart, K., Ed.; Coop. Ext. Serv., Univ. of Idaho: Moscow, 1986; pp 6–16.

APPENDIX–I   A TYPICAL MUSTARD FIELD AND CANOLA OIL.

CHAPTER 11

ENERGY ANALYSIS OF WHEAT UNDER PRESSURIZED IRRIGATION

MOHAMED E. EL-HAGAREY[1], HANI A. A. MANSOUR[2], and M. S. GABALLAH[2]

[1]Irrigation and Drainage unit, Division of Water Resources and Desert Land, Desert Research Center (DRC), Ministry of Agriculture and Land Reclamation, 1 Mathaf El-Mataria street, El-Mataria, Cairo, Egypt

[2]Water Relations Field Irrigation Department, Agricultural and Biological Division, National Research Center, Cairo, Egypt

CONTENTS

ABSTRACT

During the year 2012–2013, the two field experiments were carried out under four pressurized irrigation systems at NRC Farm, Nubaria, Beheira Governorate, This chapter investigates the energy feasibility (a net-back, revenues and applied energy analysis) of wheat (*Triticum aestivum* L. cv. Gemmaiza-9) crop under various pressurized irrigation systems: surface drip (SD), subsurface drip (buried hoses, BD), fixed sprinkler (FS), semi-portable sprinkler (PS). Applied irrigation water amounts were 50, 75, and 100% of calculated applied water (W1, W2, and W3, respectively). The experimental design was complete split plot.

The highest energy efficiency of crop irrigation (EECI) were (PS, W3), (PS, W2), (FS, W3), and (FS, W2), respectively, while the other treatment was semiclose. The highest pumping power was for FS, PS, SD, and BD irrigation systems, respectively. The highest energy requirements were (SD, W3), (SD, W2), (SD, W1), (FS, W3), and (PS, W3). The highest applied installation energy was for BD, FS, SD, and PS irrigation systems respectively. It can be observed that the highest applied operating energy was for (PS, W3), (BD, W3), (BD, W2), (PS, W2) and (SD, W3), respectively. The highest annual total irrigation energy inputs (ATEI) were for (BD, W3), (BD, W2), (BD, W1), (FS, W3), (FS, W2), (FS, W1). The highest energy-applied efficiency (EAE) was for (BD, W1), (BD, W2), (BD, W3), (SD, W1), (SD, W2), (PS, W2). The highest values of both of annual total irrigation energy inputs for applying water (ATEI) and RCE were for (BD, W1), (SD, W1), (FS, W1), (PS, W1), (BD, W2), and (SD, W2). The EECI and EP increased in the beginning for BD, SD, and FS reaching to values for PS irrigation systems.

11.1   INTRODUCTION

Energy is a fundamental factor in the process of economic agricultural development, as it provides all important services that maintain economic activity and the quality of daily life. Modern farming has become very energy intensive. Energy in agriculture plays a significant role in crop production and agro-processing [28].

This study discusses research results: to determine energy consumption and energy indexes in peach production, to investigate the efficiency of energy consumption, and to make an economic analysis of peach orchards [53].

Irrigation cost of production unit for the different water treatments was lower under the surface drip and subsurface drip irrigation systems, and it was twice under subsurface drip and surface drip irrigation systems (SD) as compared to subsurface micro drip and surface micro drip systems [14]. Basic resources of water and energy are compulsory. Therefore, a thorough overview of how water and energy are used in a PWS is necessary to identify, where pouches of energy savings are located. This general picture of the energy requirements, summarized by adequate performance indicators, will give a precise idea: (a) about the use of the energy in a PWS, (b) about how much improvements are possible, and (c) about the actions to improve the situation. A new protocol thinks globally and acts locally, and it is a proposed strategy [6]. A new pottery dripper was developed from biomaterial using local, environmental, and cheap materials and working under low-head pressure; and it resulted in low operating pressure and applied energy [15]. The peaches production unit of irrigation cost was higher under subsurface drip and SD, and it was twice as compared with subsurface micro drip and surface micro drip [36].

Closed circuits of drip irrigation system require about half of the water needed by a sprinkler or surface irrigation. Lower operating pressures and flow rates result in reduced energy costs [31]. To meet the growing demand for food, more than half of world cereal production is anticipated to be produced using irrigation by 2050 [43]. Demand for food crops has been increasing in response to number of factors including a growing global population, expanding economies in developing countries, and rising biofuels production among other factors [50]. When water is inexpensive or free, farmers make irrigation decisions based on water needs and the energy cost of pumping water, not the price of water [19, 34]. The high energy costs cause the breakeven price of corn to increase [19, 33, 34, 38, 44, 45]. Several studies analyzed the feasibility of investing in irrigation systems at the farm level [8, 20, 29, 35, 38, 45, 46]. These studies, however, focus on arid regions where water is scarce and irrigation is vital for crop production. The aforementioned analysis is insightful for arid regions because they demonstrate methods to reduce irrigation costs. However, water is relatively cheap and abundant in the southeastern United States, another humid region, and producers have little incentive to conserve water or increase water use efficiency [29, 52]. There are many studies that seek to quantify the energy consumption associated with crop production in various countries [2, 7, 9, 10, 17, 18, 21, 25, 37, 40, 48, 51]. Therefore, these studies provide little insight into the profitability of irrigating crops in humid regions such as the southeastern United States, simulated yields for irrigating corn in Iowa, and calculated the breakeven corn price for irrigation on a 52 ha field. The breakeven corn price for irrigation was \$182.18 Mg^{-1}. Irrigation was not profitable since the average price of corn used to calculate net returns was \$79 Mg^{-1}(\$2 bu^{-1}). The use of energy cost of pumping water is a proxy for the price of water [11]. Energy cost slightly influenced water demand, but crop prices had the greatest influence on irrigation water demand. Other economic research on irrigation in humid regions has primarily focused on production risk management [19], and determination of optimal irrigation scheduling to maximize net returns [5].

This chapter presents research studies on: net-back, revenues, and applied energy analysis of irrigated wheat using pressurized irrigation systems under environmental desert multi-criteria energy; and efficiencies to determine the economic impact that is related to pressurized irrigation, operating head, labor, installation, maintenance, and repairs.

11.2 MATERIALS AND METHODS

11.2.1 EXPERIMENTAL DESIGN

During 2012–2013, the two field experiments were conducted under four pressurized irrigation systems at NRC Farm, Nubaria, Beheira Governorate. The site is located at 30°31′44″ to 30°36′44″N latitude and 30°20′19″ to 30°26′50″E longitude. At the site, soil texture is sandy loam, poor in organic matter (1.3%) and $CaCO_3$ (3.8%). In addition to the soil reaction (pH of 8.2), the soil is non-saline (2.6 dSm^{-1} of the extracted soil paste). Soil water content at field capacity and wilting point were 12.6 and 4.7% on a weight basis [26]. This study aims to investigate the energy feasibility (a net-back, revenues and applied energy analysis) of cultivating wheat (*Triticum aestivum* L. cv. Gemmaiza-9), under various pressurized irriga-tion systems : surface drip (SD), subsurface drip (buried hoses, BD), fixed sprinkler (FS), and semi-portable sprinkler (PS). Applied irrigation water amounts were:

$W1$ = 50% of calculated applied water,

$W2$ = 75% of calculated applied water, and

$W3$ = 100% of calculated applied water.

The statistical design was a split plot two factorial, where the main factor was irrigation systems and the sub-main factor was applied water amounts.

The amounts of fertilizers were applied according to the recommendations of the Field Crop Institute, ARC, Egypt, Ministry of Agricultural and Land Reclamation for wheat crop. Farmyard manure (FYM) at the rate of 24 m³/ha was thoroughly mixed with 0–30 cm of the soil layer before planting. In addition, 240 kg of superphosphate per hectare (ha) (15.5% P_2O_5) and 120 kg potassium sulfate (48% K_2O) were used. Recommended dose of nitrogen (100 kg of N/ha) in two equal doses was applied at 4 and 10 weeks after germination of seeds. Wheat was sown on November 10, 2012.

11.2.2 IRRIGATION SYSTEMS

Drip irrigation system (SD and subsurface drip irrigation) consisted of centrifugal pump (35 m lift and 27 m³/h discharge) driven by a diesel engine, pressure gages, control valves, flow valves, and water source from an aquifer, main line then lateral lines, and dripper lines. For traditional drip irrigation, Gr drippers (4 lph discharge, three emitters per meter) were used. The plant-to-plant spacing was 25 cm. Gr dripper spacing was 0.3 m. The first position of drip hose was SD and the second position was subsurface drip (BD) at a depth 20 cm.

Sprinkler irrigation system (fixed sprinkler FS, semi-portable sprinkler PS**)** consisted of centrifugal pump (65 m lift and 60 m³/h discharge) driven by a diesel engine, pressure gages, control valves, flow valves, water source from an aquifer, and main line. The components of semi-portable sprinkler system (PS) were main/sub-mains and aluminum lateral pipes (inside diameters of 150, 110, and 90 mm), couplers, sprinkler head (1 m³/h) at a spacing of 12 × 12 m², and other accessories (valves, bends, plugs, and risers). The fixed sprinkler systems (FS) was similar to the portable system except that the location of water source and pumping plant were fixed.

11.2.3 IRRIGATION REQUIREMENTS

Irrigation water requirements for wheat were calculated using the climatic data at the local weather station data at Al-Beheira Governorate, the Central Laboratory for Agricultural Climate (CLAC), Ministry of Agriculture and Land Reclamation. Irrigation requirements for wheat (mm/day) were determined by calculating crop consumptive use [1, 12] and using Eq (11.1)[24]. The values are given in Table 11.1.

TABLE 11.1   Irrigation Requirements of Wheat at Nubaria Site, Egypt.

Growth stage	Month	Et_o	K_C	Et_c	Id
		mm/day		Mm/day	m³/(ha.day)
Planting	December	2.8	0.4	1.1	11.2
Rapid	January	6.3	0.4	2.5	25.2
Vegetative	February	5.9	0.8	4.7	47.2
Flowering & seed fill	March	4.2	1.3	5.5	54.6
Maturity	April	7.4	0.5	3.7	37
Harvesting	May	2.0	0.4	0.8	8
Total (Iy)	5547.8 m³ per (ha.season)				

Id = Irrigation requirements of wheat, **m³/(ha.day)**.
Iy = Seasonal irrigation requirements for wheat, **m³/(ha.season)**

$$IR = \{[K_c \times Et_o \times A \times C_F]/[10^7 \times E_a]\} + LR \qquad (11.1)$$

where A = Area irrigated in m², E_a = Application efficiency in % = 90%, LR = Leaching requirements, IR= Irrigation water requirements in m³/(ha. day), Et_o = Potential evapotranspiration in mm/day, K_C = Crop coefficient of wheat, C_F = Crop cover, Et_c = Crop evapotranspiration using K_c [1].

11.2.4   ENERGY ANALYSIS

11.2.4.1   TOTAL ENERGY INPUTS FOR IRRIGATION OPERATION

The total energy inputs for irrigation were determined in terms of per year, per unit area and volume of applied water. The total seasonal energy is the sum of the seasonal fixed installation energy, the seasonal operation energy (pumping plus maintenance), and human labor [13]. The seasonal fixed installation energy is the energy required to install the irrigation system for a useful life for the length of any evaluation period divided by the number of years of the period. In this study, the evaluation period was 20 years. Energy associated with transporting of different components to the site was not considered in this study, because of unreliable data records. The procedure to calculate total energy is described below:

a.   Installation energy (IE)
The annual fixed energy (AFE) is defined as the energy to manufacture a limited number of products used in irrigation system [3].

$$AFE = [(ERM + ERC) \times NTR] \div [ESL] \qquad (11.2)$$

where AFE = Annual fixed energy in MJ/(kg.year), ERM = The energy input to manufacture products from raw materials in MJ/kg, ERC = The energy input to manufacture products from recycled materials in MJ/kg, NTR = Number of times a product is replaced over the expected life of the system, and ESL = Expected system life in years.

The manufacturing energy for certain products used in irrigation systems has been calculated in the past [3]. The energy required for manufacturing equipment or machinery (ME), which used in excavation and land forming, was computed as follows [4, 16]:

$$ME = [kW \times 14.88 \text{ MJ/kW} + \text{Equip. wt.} \times 71.2 \text{ MJ/ton}] \times [\text{work hours/useful life in h}]$$

$$\text{Expected life (hours)} = \text{Expected life (years)} \times$$
$$\text{Activity (hours per year)} \times \text{Load factor} \qquad (11.3)$$

where ME = Manufacture energy, kW = Engine power in kW, and Equip. wt. = Weight of operating machine in ton.

The energy associated with fuel consumption was computed on the basis of 41.06 MJ/L, [39]. Energy associated with the repairs and maintenance of the machinery was estimated as 5% of machinery energy inputs [18].

Human labor energy was estimated as follows [23]:

$$EHL = [CHL/F_c] \times NL \qquad (11.4)$$

where EHL = Human labor energy in MJ/ha, CHL = Energy input coefficient represents the human labor energy = 2.3 MJ/human, NL = The number of persons required for any operation, and F_c = Field capacity, ha/h.

b. Operation energy (OE)

Energy inputs in the operation for irrigation system include maintenance and pumping energies (PEs). Annual maintenance energy for irrigation system was roughly estimated as 3% of annual installation energy [3]. The PE was calculated as follows [3, 30]:

$$PE = [K]\{[A \times D \times H]/[E_p \times E_i]\} \qquad (11.5)$$

where PE = Pumping energy in MJ/ha, K = Conversion factor depending on the units used, A = Area irrigated in hectares, D = Net depth of irrigation water requirement in m, H = Pumping head in m, E_p = Pumping system efficiency, and E_i = Irrigation efficiency.

c. Human labor energy (EHL)
The energy associated with labor for system operation and management was determined as follows [3]:

$$\text{EHL} = \{[t \times n \times c]/[A]\} \times \text{NL} \qquad (11.6)$$

where EHL = Human labor energy in MJ/(ha.year), t = Time for one irrigation in h, n = Number of irrigations in the year, c = Energy input coefficient representing human labor energy = 1.26 MJ/(man-hours), NL = The number of persons required for one irrigation, and A = Area irrigated in ha.

Human labor energy inputs associated with the operation and control of the water in this study were those of manual man-hours with water control structures installed; and is negligible energy input of less than 0.42 MJ/(ha.year) [13].

11.2.4.2   ENERGY YIELD [7, 41]

a.   Relative consumed energy (RCE)

$$\text{RCE, (MJ/kg)} = \text{Total energy input (MJ/ha)} \div \text{wheat yield (kg/ha)} \qquad (11.7)$$

b.   Energy efficiency of irrigation (EECI) (energy ratio, %)

$$\text{Energy ratio} = \text{Total energy outputs (MJ/ha)} \div \text{total energy inputs (MJ/ha)} \qquad (11.8)$$

c.   Annual total irrigation energy outputs (ATEO) in MJ/(ha.year)

$$\text{ATEO} = \text{Wheat grain yield (kg/ha)} \times \text{the digestible energy of wheat (kg)} \qquad (11.9)$$

The digestible energy of wheat = 16.4 MJ/kg [From 46].

$$\text{Energy productivity} = \text{Wheat yield (kg/ha)} \div \text{total input energy (MJ/ha)} \qquad (11.10)$$

$$\text{Net energy gain (MJ/ha)} = \text{Total energy output (MJ/ha)} - \text{Total energy input (MJ/ha)} \qquad (11.11)$$

d. Energy requirements and energy-applied efficiency (EAE) [3]
    Power consumption use for pumping water (BP):

$$BP = [Q \times TDH \times Yw]/[E_i \times E_p \times 1000] \qquad (11.12)$$

where Q = Total system flow rate in m³, TDH = Total dynamic head in m, E_i = Total system efficiency, E_p = Pump efficiency, and Yw = Water specific weight = 9810 N/m³.

Pumping energy requirements, Er, in kW.h Er = BP $\times H$ (11.13)

where H = Irrigation time per season in hours.
    Pumping energy applied efficiency (EAE)

$$EAE, (kg./(kW.h)) = [\text{Total fresh yield in kg}] \div [\text{Energy requirements in kW.h}] \qquad (11.14)$$

11.3 RESULTS AND DISCUSSION

The data on energy analysis is shown in Table 11.2, and Figures 11.1–11.4. The highest irrigation pumping power consumption was for FS, PS, SD, and BD irrigation systems, respectively. The highest energy requirement was for combinations (SD, W3), (SD, W2), (SD, W1), (FS, W3), and (PS, W3) and the other treatments were semi-close. The highest applied installation energy was for BD, FS, SD, and PS irrigation systems, respectively, as can be observed that the last energy parameters lead to the operating and annual total energy. It is concluded that the highest applied operating energy was for treatments: (PS, W3), (BD, W3), (BD, W2), (PS, W2), and (SD, W3), respectively. The highest ATEI was for treatments: (BD, W3), (BD, W2), (BD, W1), (FS, W3), (FS, W2), and (FS, W1). The type of irrigation system has an impact on the amount of energy consumed, even within pressurized systems, because the energy required for pumping depends on the total dynamic head, flow rate, and system efficiency [27].

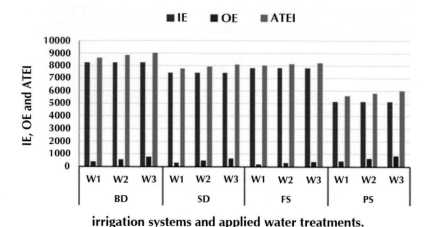

FIGURE 11.1   The installation energy inputs (IE), operating energy (OE), annual total energy inputs (ATEI), versus irrigation systems and applied water treatments.

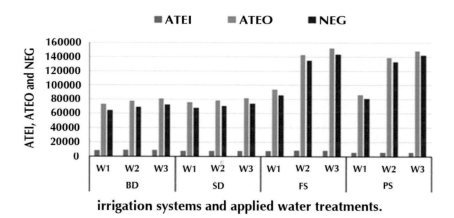

FIGURE 11.2   The annual total energy inputs (ATEI), annual total energy outputs (ATEO), and net energy gain (NEG) versus irrigation systems and applied water treatments.

Indirect irrigation, energy inputs are associated with the energy embodied in irrigation infrastructure and its operation [22, 27]. The approximately 23% of direct energy use in crop production was used for on-farm pumping. Subsequently, the highest ATEO and NEG were for treatments: (FS, W3), (PS, W3), (FS, W2), (PS, W2), and the other treatments were semi-close. The significant differences were between the ATEO and ATEI. The highest ATEI was for (BD, W3), (BD, W2), (BD, W1), (FS, W3), (FS, W2), (SD, W1), and (FS, W1); and the lowest ATEI was for (PS, W1). The highest EAE was for (BD, W1), (BD, W2), (BD, W3), (SD, W1), (SD, W2), and (PS, W2), while the other treatments were not far away about the last value. The highest value of both of AIEI and RCE was for (BD, W1), (SD, W1), (FS, W1), (PS, W1), (BD, W2), and (SD, W2). Conversely, both EECI and EP were increased in the beginning for BD, SD, FS, and reaching to values for PS irrigation systems.

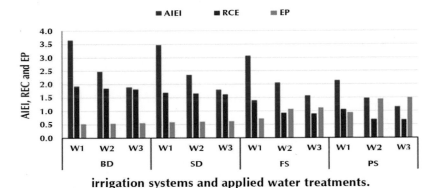

irrigation systems and applied water treatments.

FIGURE 11.3   The annual total irrigation energy inputs for applying water (AIEI), relative consumed energy (RCE), and energy productivity (EP) versus irrigation systems and applied water treatments.

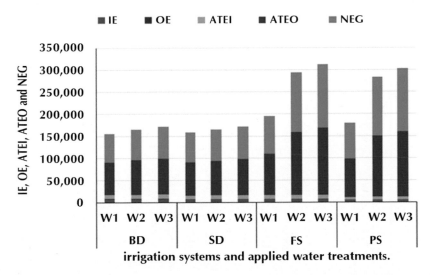

irrigation systems and applied water treatments.

FIGURE 11.4   The installing energy inputs (IE), operating energy (OE), annual total energy inputs (ATEI), annual total energy outputs (ATEO), and net energy gain (NEG) versus irrigation systems and applied water treatments.

TABLE 11.2 Energy Feasibility Analysis of Pressurized Irrigation Systems and Applied Water Amounts.

Item	Irrigation system, IS											
	BD			SD			FS			PS		
	W1	W2	W3	W1	W2	W3	W1	W2	W3	W1	W2	W3
	Applied water amounts, WA											
BP	2.4	2.4	2.4	4.3	4.3	4.3	17.7	17.7	17.7	12.5	12.5	12.5
IE		8306			7466			7848			5172	
OE	376	564	751	329	494	658	197	295	393	428	642	856
ATEI	8682	8870	9057	7795	7960	8124	8045	8143	8241	5600	5814	6028
AWU	2383	3574	4766	2247	3371	4494	2622	3933	5244	2621	3931	5244
AIEI	3.6	2.5	1.9	3.5	2.4	1.8	3.1	2.1	1.6	2.1	1.5	1.1
ER	229	343	458	1714	2572	3429	465	697	929	393	589	786
EAE	20	14	11	13	9	7	7	8	6	8	9	7
Yield	4488	4769	4968	4622	4786	5006	5729	8717	9259	5297	8450	9041
RCE	1.93	1.86	1.82	1.69	1.66	1.62	1.40	0.93	0.89	1.06	0.69	0.67
ATEO	73,603	78,208	81,475	75,807	78,484	82,105	93,952	142,956	151,851	86,868	138,587	148,269
EECI	8.48	8.82	9.00	9.72	9.86	10.11	11.68	17.56	18.43	15.51	23.84	24.60
EP	0.52	0.54	0.55	0.59	0.60	0.62	0.71	1.07	1.12	0.95	1.45	1.50
NEG	64,921	69,339	72,418	68,012	70,524	73,981	85,908	134,812	143,609	81,268	132,773	142,241

The statistical analysis indicated significant impact of both of applied water amounts and irrigation systems on the energy parameters. There was significant influence of the applied water amounts, the pressurized irrigation systems, and the interaction of these on all of the energy parameters. The interaction impact was also significant in IE, ATEI, AIEI, and RCE at LSD = 0.05.

The significant difference in pumping power of sprinkler irrigation system and drip irrigation system is due to higher operating head, which is necessary to operate sprinkler water jet. The pumping power of FS was higher than the PS irrigation system, due to the number of operating sprinklers. It can be observed that the number of operating sprinklers in 1 ha is 70 sprinklers in FS irrigation systems. In comparison, the number of operating sprinklers in 1 ha is 24 sprinklers in the PS irrigation systems. This difference attributes to the significant differences in the applied pumping power [27, 47] among FS and PS. In addition, the total operating head of FS is higher than the PS irrigation systems. The higher human labor energy is needed for portable sprinkler irrigation systems [32, 39].

In Table 11.3, it can be noted that the highest installation energy is for BD, FS, SD, and PS irrigation systems, due to annual fixed energy that is related to the weight of material of irrigation system per hectare. The weights of both of PVC and PE of BD, SD, FS, and PS are 195, 195, 1278, and 876 kg of PVC per hectare, and 250, 250, 5.5, and 1.92 kg of PE per hectare. The highest manufacture energy of FS, PS, BD, and SD irrigation systems is 10.6, 7.2, 4.6, and 3.92 MJ/ha, respectively. It is related to the excavation and backfill cubes of soil to install the irrigation systems besides the ratio of work capacity [22, 27]. We should also consider the number of man hours for a specific job per hectare. The highest human labor energy per hectare was for (PS, W3), (BD, W3), (SD, W3), and (FS, W3), respectively. For applied water amounts, the highest human labor energy per hectare was for W3, W2, and W1, respectively [32].

The lowest value of both of AIEI and RCE was for PS, FS, SD, and BD, respectively; and for W3, W2, and W1 according to the applied amounts of water. We need more applied energy for pumping. Correspondingly, the highest values of EP, ATEO, and NEG were for FS, PS, SD, and BD due to the highest grain yield of wheat in sprinkler systems, compared to drip irrigation systems [42].

Finally, it can be noted that the means NEG of (FS, W3), (PS, W3), (FS, W2), and (PS, W2) are higher than the means of the other treatments by 47% approximately. The highest EECI was for (PS, W3), (PS, W2), (FS, W3), and (FS, W2), respectively. The highest overlap irrigated area is for sprinkler

TABLE 11.3 The Influence of Pressurized Irrigation Systems and Water Amounts of the Various Energy Parameters.

IS	WA	IE	OE	ATEI	AWU	AIEI	Er	EAE	RCE	ATED	EECT	EP	NEG
BD	W_1	8306	376a	8682a	2383a	3.6a	229a	20a	1.93a	73603a	8.48a	0.52a	64921a
	W_2		564b	8870b	3574b	2.5b	343b	14b	1.86ba	78208b	8.82ba	0.54ba	69339b
	W_3		751c	9057c	4766c	1.9c	458c	11c	1.82cab	81475c	9.00cab	0.55cab	72418c
Mean			564b	8870a	3574d	2.7a	343d	15a	1.87a	77762d	8.77d	0.54dc	68893d
SD	W_1	7466	329a	7795a	2247a	3.5a	1714a	13a	1.69a	75807a	9.72a	0.59a	68012a
	W_2		494b	7960b	3371b	2.4b	2572b	9b	1.66ba	78484b	9.86ba	0.60ba	70524b
	W_3		658c	8124c	4494c	1.8c	3429c	7c	1.62cab	82105c	10.11cab	0.62cba	73981c
Mean			494cb	7960c	3371c	2.6ba	2572a	10b	1.66b	78799c	9.90c	0.60c	70839c
FS	W_1		197a	8045a	2622a	3.1a	465a	7ba	1.40a	93952a	11.68a	0.71a	85908a
	W_2	7848	295b	8143b	3933b	2.1b	697b	8a	0.93b	142956b	17.56b	1.07b	134812b
	W_3		393c	8241c	5244c	1.6c	929c	6cb	0.89cb	151851c	18.43c	1.12cb	143609c
Mean			295d	8143b	3933a	2.3c	697b	7dc	1.07c	129586a	15.89b	0.97b	121443a
PS	W_1		428a	5600a	2621a	2.1a	393a	8ba	1.06a	86868a	15.51a	0.95a	81268a
	W_2	5172	642b	5814b	3931b	1.5b	589b	9a	0.69b	138587b	23.84b	1.45b	132773b
	W_2		856c	6028c	5244c	1.1c	786c	7cb	0.67ba	148269c	2460cb	1.50cb	142241c
Mean			642a	5814d	3932ba	1.6d	589c	8c	0.81d	124575b	21.32a	1.30a	118761b
LSD 0.05			68	165	408	0.3	88	1.5	0.16	1578	3.7	0.22	1726
Mean	W_1		332.5a	7530.5a	2468a	3.1a	700.3a	12a	1.52a	82558a	11.35a	0.69a	75027a
	W_2		498.8b	7696.8b	3702b	2.1b	1050.3b	10b	1.29b	109559b	15.02b	0.92b	91558b
	W_3		664.5c	7862.5c	4937c	1.6c	1400.5c	8c	1.25cb	115925c	15.54cb	0.95cb	108062c
LSD 0.05			102	117	756	0.8	347	1.1	0.12	2976	2.4	0.12	1236
$LSD_{0.5}$ ($I \times II$)			23	42	146	0.4	104	0.2	0.02	519	1.2	0.06	452

systems in comparison with drip irrigation. We need higher land area for surface drip tubes to cover the intensive cultivated area by wheat.

11.4 CONCLUSIONS

The sprinkler irrigation systems have higher net-back energy compared to drip irrigation system for wheat cultivation. The sprinkler irrigation systems need more total operating head. However, one should also consider high operating hours of irrigation process, the plant intensity, and the covering efficiency of applied water under drip irrigation compared to any type of sprinkler irrigation systems [27, 49].

ACKNOWLEDGMENT

Authors are grateful for the financial support from the National Research Center (NRC), and the Desert Research Center, DRC, Cairo, Egypt: The Special Fund for Agro-Scientific Research in the Public Interest (Balanced fertilization of main crops in Egypt).

KEYWORDS

- annual total irrigation energy inputs
- applied energy
- applied installation energy
- desert
- drip irrigation
- economy
- Egypt
- energy
- energy efficiency
- fixed sprinkler
- installation energy
- pressurized irrigation
- pumping

- **pumping power requirements**
- **semi-portable sprinkler irrigation**
- **sprinkler irrigation**
- **subsurface drip irrigation**
- **surface drip irrigation**
- **water**
- **wheat**

REFERENCES

1. Allen R. G.; Pereira, L. S.; Raes, D.; Smith, M. *Crop Evapotranspiration - Guidelines for Computing Crop Water Requirements;* FAO Irrigation and Drainage FAO - Food and Agriculture Organization of the United Nations: Rome, 1998; p 56.
2. Barber, A. *Seven Case Study Farms: Total Energy and Carbon Indicators for New Zealand Arable and Outdoor Vegetable Production*; Agrilink: New Zealand, 2004; pp 224.
3. Batty, J. C.; Keller, J. Energy Requirement for Irrigation. In *Handbook of Energy Utilization in Agriculture;* D. Pimentel, Ed.; CRC Press: Florida, USA, 1980; pp 35–44.
4. Batty, J. C.; Hamad, S. N.; Keller, J. Energy Inputs to Irrigation. *J. Irri. Drain. Div. ASCE.* **1975,** *101*(IR4), 293–307.
5. Boggess, W. G.; Lynne, G. D.; Jones, J. W.; Swaney, D. P. Risk-Return Assessment of Irrigation decisions in Humid Regions. *Southern J. Agric. Econ.* **1983,** *15,* 135–143.
6. Cabrera, E.; Cabrera, E.; Cobacho, R.; Soriano, J. Towards an Energy Labelling of Pressurized Water Networks. *Procedia. Eng.* **2014,** *10,* 209–217.
7. Canakci, M.; Topakci, M.; Akinci, I.; Ozmerzi, A. Energy Use Pattern of Some Field Crops and Vegetable Production: Case Study for Antalya Region, Turkey. *Energy Convers. Manage.* **2005,** *46*(4), 655–666.
8. Caswell, M.; Zilberman, D. The Effects of Well Depth and Land Quality on the Choice of Irrigation Technology. *Am. J. Agric. Econ.* **1986,** *68,* 798–811.
9. Chamsing, A.; Salokhe, V.; Singh, G. Energy Consumption Analysis for Selected Crops in Different Regions of Thailand. *Agric. Eng. Int. CIGRE. e-J.* **2006,** *8,* Paper # 06-013, 1–18.
10. Chaudhary, V.; Gangwar, B.; Pandey, D. Auditing of Energy Use and Output of Different Cropping Systems in India. *Agric. Eng. Int. CIGRE. e-J.* **2006,** *8,* Paper # 05-001, 1–13.
11. Dejonge, K. C.; Kaleita, A. L.; Thorp, K. R. Simulating the Effects of Spatially Variable Irrigation on Corn Yields, Costs, and Revenue in Iowa. *Agric. Watermanage.* **2007,** *92,* 99–109.
12. Doorenobs, J.; Pruitt, W. O. *Guidelines for Predicting Crop Water Requirements*; FAO Irrigation And Drainage Paper 24: Rome, Italy, **1977**; pp 156.

13. Down, M. J.; A. K. Turner; T. A. Mcmahon. *On Farm Energy in Irrigation;* Final Report No. 78/86 of a Project Supported by the NER. Development and Demonstration Council. Melbourne Univ: Civil and Agric. Eng. Dept:, Australia, 1986; pp78 .

14. El-Hagarey M. E.; El-Nesr, M. N.; Mehanna, H. M.; Mansour, H. A.. Energy, Economic Analysis and Efficiencies of Micro Drip Irrigation I- Energy Analysis. *IOSR-JAVS.* **2014,** *7*(8), 19–26.

15. El-Hagarey, M. E.. Design and Manufacture of Pottery Dripper for the Use of Saline Water in Irrigation Systems. *IOSR-JAVS.* **2014,** *7*(5), 70–80.

16. EPA (Assessment And Standards Division) *Median Life, Annual Activity, And Load Factor Values For Nonroad Engine Emissions Modeling*; Report No. NR-005b, **2002,** pp1–47.

17. Erdal, G.; Esungun, K.; Erdal, H.; Gunduz, O. Energy Use and Economic Analysis of Sugar Beet Production in Tokat Province of Turkey. *Energ.* **2007,** *32,* 35–41.

18. Esungun, K.; Erdal, G.; Gunduz, O.; Anderdal, H. An Economic Analysis and Energy Use in Staked Tomato Production in Tokat Province of Turkey. *Renew. Energ.* **2007,** *32,* 1873–1881.

19. Gonzalez-Alvarez, Y.; Keeler, A. G.; Mullen, J. D. Farm-Level Irrigation and the Marginal Cost of Water Use: Evidence from Georgia. *J. Environ. Manage.* **2006,** *80,* 311–317.

20. Guerrero, B. L.; Amosson, S. H.; Marek, T. H.; Johnson, J. W. Economic Evaluation of Wind Energy as an alternative to Natural Gas Powered Irrigation. *J. Agric. Appl. Econ.* **2010,** *42,* 277–287.

21. Hatirli, S. A.; Ozkan, B.; Fert, C. Energy Inputs and Crop Yield Relationship in Greenhouse Tomato Production. *Renew. Energ.* **2006,** *31,* 427–438.

22. Hodges, A. W.; Lynne, G. D.; Rahmani, M.; Casey, C. F. *Adoption of Energy and Water-Conserving Irrigation Technologies In Florida.* University Of Florida: USA, 1994.

23. Kassem, A. S. A Mathematical Model for Determining Total Energy Consumption for Agriculture Systems. *Misr. J. Agric. Eng.* **1986,** *3*(1), 39–57.

24. Keller, J.; D. Karmeli. *Trickle Irrigation Design;* Rain Bird Sprinkler Co., Glendor, CA, ASABE 91740, USA, 1975; pp 24–26.

25. Khan, S.; Khan, M. A.; Hanjra, M. A.; Mu, J. Pathways to Reduce the Environmental Footprints of Water and Energy Inputs in Food Production. *Food Policy.* **2008,** *34* (2), 141–149.

26. Klute, A.. Water Retention: Laboratory Methods. In *Methods Of Soil Analysis, Part I - Physical And Mineralogical Methods ; A. Klute, Ed.;* Monograph 9 by ASA and SSSA: Madison, WI, 1986; pp 635–662.

27. Lal, R.. Carbon Emission from Farm Operations. *Environ. Int.* **2004,** *30,* 981–990.

28. Larson, D. L.; D. D. Fangmeier, D. D. Energy in Irrigated Crop Production. *J. ASAE.* **1978,** *21,* 1075–1080.

29. Letey, J.; Dinar, A.; Woodring, C.; Oster, J. D. An Economic Analysis of Irrigation Systems. *Irrig. Sci.* **1990,** *11,* 37–43.

30. Lsraelson, O. W.; V. E. Hansen, Eds. Flow of Water Into and through Soils. In *Handbook of Irrigation Principal and Practices*; 3rd ed.; John Wiley and Sons, Inc: New York, USA; 1962; 235–240.

31. Mansour, H. A.; Mehanna, H. M. ; M. E. El-Hagarey; Hassan, A. S. Using Automation Controller System and Simulation Program for Testing Closed Circuits of Mini-Sprinkler Irrigation System. *OJMSI.* **2013,** *1*(2), 14–23.

32. Martin, J. F.; Diemont, S. A. W.; Powell, E.; Stanton, M.; Levy-Tacher, S. Energy Evaluation of the Performance and Sustainability of Three Agricultural Systems with Different Scales and Management. *Agric. ecosyst. Environ.* **2006,** *115*(1–4), 128–140.

33. Mcallister, A. A Study of the Sequential Digestion of Barley, Corn, Sorghum and Wheat by Rumen Microorganisms. *Can. J. Anim. Sci.* **1990,** *70,* 571–579.

34. Mullen, J. D.; Yu, Y.; Hoogenboom, G. Estimating the Demand for Irrigation Water in a Humid Climate: A Case Study from the Southeastern United States. *Agric. Water Manage.* **2009,** *96,* 1421–1428.

35. O'Brien, D. M.; Lamm, F. R.; Stone, L. R.; Rogers, D. H. Corn-Yield and Profitability for Low-Capacity Irrigation Systems. *Appl. Eng. Agric.* **2001,** *17,* 315–321.

36. Omima M. E. ; M. E. El-Hagarey. Evaluation of Ultra-Low Drip Irrigation and Relationship between Moisture and Salts in Soil and Peach (*Prunsperssica*) Yield. *J. Am. Sci.* **2014,** *10*(8), 13–28.

37. Ozkan, B.; Fert, C. ; Andkaradeniz, C. F. Energy and Cost Analysis for Greenhouse and Open-Field Grape Production. *Energ.* **2007,** *32,* 1500–1504.

38. Peterson, J. M.; Ding, Y. Economic Adjustments to Groundwater Depletion in the High Plains: Do Water-Saving Irrigation Systems Save Water. *Am. J. Agric. Econ.* **2005,** *87,* 147–159.

39. Phocaides, A. *Handbook on Pressurized Irrigation Techniques.* Rome: Food and Agriculture Organization of the United Nations.

40. Pimental, D.; Doughty, R.; Carothers, C.; Lamberson, S.; Bora, N.; Lee, K. Energy Inputs in Crop Production in Developing and Developed Countries. In *Food Security And Environmental Quality In The Developing World; R. Lal,* Ed.;, CRC Press: USA, **2002**; pp 129–151.

41. Rao, A. R.; Malik, R. K. Methodological Considerations of Irrigation Energetics. *Energ.* **1982,** *7*(10), 855–859.

42. Rodríguez – Díaz, D. J. A.; Pérez-Urrestaraz, L.; Camacho- Poyato, E.; Montesinos, P. The Paradox of Irrigation Scheme Modernization: More Efficient Water Use Linked to Higher Energy. *Span. J. Agric. Res.* **2011,** *9*(4), 1000–1008.

43. Rosegrant, M. W.; Ringler, C.; Zhu, T. Water For agriculture: Maintaining Food Security under Growing scarcity. *Annu. Rev. Environ. Resour.* **2009,** *34,* 205–222.

44. Scheierling, S. M.; Loomis, J. B.; Young, R. A. Irrigation Water Demand: A Meta-Analysis of Price Elasticities. *Water Resour. Res.* **2006.** Http://Dx.Doi.Org/10.1029/2005WR004009.

45. Seo, S.; Segarra, E.; Mitchell, P. D.; Leatham, D. J. Irrigation Technology Adoption and its implications for Water Conservation in the Texas High Plains: A Real Option Approach. *Agric. Econ.* **2008,** *38,* 47–55.

46. Sheriff, G. Efficient Waste? Why Farmers over-apply Nutrients and the Implications for Policy Design. *Rev. Agric. Econ.* **2005,** *24,* 542–557.

47. Singh, H.; Mishra, D.; Nahar, N. M. Energy Use Pattern In Production Agriculture Of A Typical Village In Arid Zone In India. *Energ. Convers. Manage.* **2002,** *43,* 2275–2286.

48. Singh, H.; Singh, A. K.; Kushwaha, H. L.; Singh, A. Energy Consumption Pattern of Wheat Production in India. *Energ.* **2007,** *32,* 1848–1854.

49. Srivastava, R. C.; Verma, H. C.; Mohanty, S.; Pattnaik, S. K. Investment Decision Model for Drip Irrigation. *Irrigation Sci.* **2003,** *22,* 79–85.

50. Trostle, R. *Global Agricultural Supply And Demand Factors Contributing To The Recent Increase In Commodity Prices.* Report WRS-0801, US Department of Agriculture, Economic Research Service: Washington DC, 2008.

51. Tzilivakis, J.; Warner, D. J.; May, M.; Lewis, K A.; Jaggard, K. An Assessment of the Energy Inputs and Greenhouse Gas Emissions in Sugar Beet (*Beta Vulgaris*) Production in The UK. *Agric. Syst.* **2005,** *85,* 101–119.
52. Vories, E. D.; Tacker, P. L.; Lancaster, S. W.; Glover, R. E. Subsurface Drip Irrigation of Corn in the United States Mid-South. *Agric. Water Manage.* **2009,** *96,* 912–916.
53. Zarini, R. L.; Akram A. Energy Consumption and Economic Analysis for Peach Production in Mazandaran Province of Iran. *The Exp.* **2014,** *20*(5), 1427–1435.

INDEX